ALONG THE TRAIL

Becoming an Israeli Commando Fighter

ALONG THE TRAIL
Becoming an Israeli Commando Fighter

ISBN# 978-1-938842-56-6

Interior layout by *Bardolf & Company.*

Cover design by Roni Ashernizky.

First published in Israel as *Kol Haderech* in 2014.

Dedicated to my friends
who are not with us,
Dany Senesh and Shai Shacham.

Blessed be their memories.

ALONG THE TRAIL

Becoming an Israeli Commando Fighter

Ouri Tsafrir

Bardolf & Company
Sarasota, Florida

Table of Contents

Introduction to the English Edition

Kol Haderech—Along the Trail—is based on my memories as a soldier in the elite military commando outfit of the Israeli Defense Force (IDF), known as "the Unit," 40 years ago. I wrote it to tell my story to myself, to my children and to others. Today, everybody talks about the Unit, but in the past its Hebrew name was a secret whispered from ear to ear: **Sayeret Matkal**.

This book describes a journey to the edge of human capability. It is a journey of determination and persistence into the hearts of Sayeret Matkal's fighters—a crew of young, dedicated men blazing their way to the highest peak of the Israeli ethos and Israeli identity. We completed extreme challenging walking tours, drilling exercises and secret operations that will stay forever behind sealed lips. Our harsh, demanding tasks and missions were designed to toughen us against ice-cold nights and searing hot days, among the sharp volcanic rocks of the Golan Heights, the freezing snows of the Hermon Mountain, the swamps of the Northern Sinai Desert towards the Suez Canal, and through the endless desert sands, deep behind enemy lines.

The crew is the basic block of the Unit, each with eight to 16 combat soldiers. During the time of my service, there were no more than six operational crews, most with 10 to 12 men. They all had their own, unique characteristics and personality, and the rivalry among them was fierce. One of the reasons we liked to call our team a "crew" was that, for us, the most important notion was, "We came here to do the tasks at the highest level of quality and professionalism in the least amount of time."

In this book, I will use *crew*, *team*, and *group* interchangeably for the sake of variety. There is a Glossary at the end to provide context and explanation for English readers unfamiliar with some of the Hebrew words, cultural expressions, and technical and military terms.

Ultimately, the book attempts to answer the most incisive questions: How do you carry on and complete your mission when your strength has been exhausted? How do you keep on marching when your legs collapse? How does a hand-picked crew become an undefeatable fighting force? What brings these soldiers together? What is the role of the commander when the crew crystallizes? What drives the best fighters in the Israeli Defense Forces? And what is their secret?

I expect that, some of the answers apply to elite military forces of other countries as well. I hope this book opens a window into understanding what we demand of our young men and women who put themselves into the line of fire to protect our country.

Writing this book was a healing journey for me and has created an opportunity to process my experiences and start a conversation with people and my IDF comrades. I don't pretend that my memories give an accurate description of reality, but they express my feelings and experiences as I remembered them. Although, memories can be deceiving.

Ouri Tsafrir
June 2022

Northern Israel after the 1973 War.

MAP OF ISRAEL

The bicycle route from Shoham to Eilat is indicated by the interrupted line.

Biking to Eilat

The Arava desert heat envelops me. I am thirsty and hungry. My legs are moving mechanically. Right forward, left forward, right, left, right, left, right, left. Breathing is steady, not labored. The wind is at my back. I straighten my back so it becomes a sail. Take advantage of the wind.

On. On. On. The word echoes in my head. The mobile phone rings again. I have no energy to answer. Let it ring. Take the water bottle out of its pocket, again. Bite down on the mouthpiece and pull it out. Take a sip. Breathe. Sip again. Hand the bottle to Yadi who rides to my left. He nods and smiles. Continuing side by side. Pushing the pedals in circles. Right, left, right, left, right, left. Accept the bottle from Yadi. Take two sips and return the bottle to its pocket.

The greyness of the asphalt has accompanied us for more than eleven hours. We ride on the right side of the yellow line, our consciousness focused on the familiar circular motion of the pedals. Right, left, right, left. We have ridden for over 125 miles. More than 40 remain. No. We are not professional cyclists. Amateurs with a dim memory of the distant past. Younger me emerges hesitantly.

Moshav* Paran is the next stop. Another 20 mile. My muscles ache. My hands cramped in pain. My ass screams. Continue. My consciousness returns to the endless rotation of the pedals. Right, left, right, left. Continue.

Too much strain. Muscle cramp. Reduce the effort a bit. The muscle relaxes. Back at the infinite circle. Right, left, right, left. The sun

* a small agricultural settlement.

13

is setting west in an abundance of desert colors. Another mile passes. I stick close to the leader, to Yadi. Moderate slope. Exceed again the limits of my endurance and the muscle cramps. Decrease the load and the muscle relaxes. A delicate balance is required to keep up with Yadi's pace and not experience muscle cramps. Having difficulty keeping the necessary focus. Count on the meditative pace. The dry heat does not exhaust but remains thick. Continue to pedal after Yadi. The right hip muscle makes its presence felt.

The physical effort is like the force exerted by the magnet of a black hole. Swirling abyss. Uncovers long forgotten personal memories. Heights of euphoria I touched in the past when my body crossed boundaries I did not believe exist. The strenuous effort melts barriers. Enables to touch my soul. Dissolves armor. Opens a window into the essence of my life. Leads to my hidden, innermost, trembling core.

Get back into the rhythm.

How did we start this ride? Why do we continue? Arrgh!

In the beginning, God created the heavens and the Earth. And for us, the sweating, the Earth.

When I was approaching the age of fifty, we decided to ride from Shoham, my town, to Eilat by bicycle.

Three in the morning. Wake up before the alarm clock. The clock starts ringing after a few minutes. Turn it off and go down to the living room.

I say "Good morning" to Yadi.

"Good morning," he replies, and begins packing.

Yadi and Amnon arrived last night. Yadi from Nahalal and Amnon from Gamla. We drink water. "Drink a lot of water" we murmur. Smiling inside, aware of our familiarity with this phrase uttered during so many events we experienced together. Wearing cycling shorts, shirt and a light windbreaker. Drink again. A little tea. Almost 4 .a.m. Ovad arrives from Paran. Addie, who flew in two days earlier from his home in Holland, comes from Herzliya. Hagai, our wounded friend from Kibbutz Ramot Menashe, comes with his companion. Ben Ami from Kochav Yair. Doron is in the USA and Avner in Brazil. Lonny of Jerusalem arrives from Bnei Zion. Enters with a smile and says, "Like in the movie *The Magnificent Seven*. Everyone comes to do the job." All smiles. Fixing the warning lights to our windbreaker, helmet and seat. Checking the tires again. Shai Porat from Moshav Bnaya, my friend, not from the IDF, arrives with a smile. He will escort us in the pickup truck. Taking photographs. Stumbling in the professional bikers' shoes on the Dutch street bricks adjacent to our apartment in Shoham. Last photo. Set off to meet Shai at the Beit Nabala intersection. We ride towards the northern gate of Shoham, which is closed now.

Before the gate we dismount from the bikes to push them towards the gate. Strange to walk in my bikers' shoes. It's hard to keep my balance. The surface is slick. I slip. Fall to my knees. "*Kus emek*,"* I curse as I stand up and bypass the gate. Inhale and exhale out a long breath. Preparing to climb clumsily on the bike's seat. Lock left foot into the pedal. Start pedaling. Lock right. Breathe. Leading the group out and to the right. The cross roads' lighting fades away. The darkness is diluted by the lights of my town on the right.

Cold. The wind meets me unprepared for the darkest hours of the night. The cold reminds my body of long cold nights as a soldier in the Golan Heights. Now, more than thirty years later, I love that cold. I can choose to leave it and climb into the pick-up, or go home if I want. Can drink warm tea as much as I want and when I want.

* an obscene curse word, originally from Arabic.

15

And this certainty fills me with joy that bursts out with a shout that wakes Yadi from his reveries. A steep slope and we are at the Beit Nabala junction. Recall the song that probably was erased from today's military culture. *Beit Nabalah, speed up, Beit Naballah, speed up.*

Meet Shai and the pickup. We take off our windbreakers and put them in the truck bed. My body starts to warm up. Everyone is still in a drowsy morning trance. Situated within the riders group. We push southward along the road. First sun light gets stronger on our left. Finding our rhythm. Not too fast or too slow. Next stop: Kastina. The road points southwest. We ride against the wind. Challenging and strenuous. The leader fights the wind, clearing the path for us.

When the leader gets tired, he signals. The second rider leads. Facing the wind. The feet are pedaling meditatively. Right, left, right, left, right, left.

Breathing is steady. I get into the rhythm.

Acceptance Tests

God heard them and tested them, ten days.
(Daniel, A, V. 14)

My IDF physical score was set at 89 because I was nearsighted. Yoram, from my Kibbutz, who was two years ahead of me in the Unit, recommended that I go to the tests without glasses. At any rate, I put the glasses on only during movies. I was ashamed. Afraid that the other kids would call me "Mishkafofer"—"glasses wearer." I left the glasses in my room in the cupboard and went to the tests feeling stressed out.

The tests took place in Kule, close to Kibbutz Nachshonim. We met under a bridge. Started by carrying a stretcher. We were boys age 17 to 18. Who knew what it takes to carry a stretcher? All we knew is that those who could lug for a longer period would score more points. The carrying twisted my body, and I did not know how to deal with that. I recalled the story of Ben-Hur who alternated rowing on the left and right sides of the Roman galley to keep his muscles in balance. It was hard. As time passed and the road continued, more people dropped out and fewer of us remained to carry the stretcher. Heavy breathing and sweating became my companions. We went a short distance that to us, the untrained, seemed endless. When we got back to the bridge, we thought we were done for the day. Dany Bruner, the testing officer who led us, acted like a tourist, strolling around. I felt exhausted. Drank much water.

As we were sitting under the bridge, I was amazed to see one of the soldiers climbing up a rope using his hands only. *How is he doing it?* I knew how to climb a rope by creating a "step" with my legs, break and squat style.*

But with no legs? Now, they asked us to climb. Hands only. I started to climb. Halfway to the top I got stuck. Could not move my left hand above my right. I hung in this position for several unbearable seconds. Then, I heard the soldier below order, "Climb with your legs." With no hesitation I used the "step" practice with the rope and slowly climbed up to the top.

After all of us climbed, came down and drank water, we were asked to climb again. Insane. But we were being tested. This time I get stuck after barely two yards up the rope. Second after second ticks away. I am stuck. My muscles start to burn. A few more seconds and I start to slide down. The examiner calls to me, "Catch with your legs." I tighten my legs and stop sliding. Then, receiving the assistance I had not asked for, I came down.

At the end of the day I was exhausted. I felt that I made a valiant effort and that the examiners saw it. I felt anxious. *Did I pass?* Mixed with the feeling: *They noticed that I am good.* My general mood was positive. I lived with complete ignorance of where I so eagerly desired to serve. I was attracted by the glory and honor of being among those who "knew." Was attracted to something I had no real idea what it was. Just heard about. And imagined. Why was I so drawn to it? To be one of "The Chosen"? To succeed in the most demanding service path?

The knowledge that I was OK felt good.

* Letting the rope fall on the outside of a leg, "stepping" on it by using the opposite foot to help. The rope falling across the top of the foot and while stepping on it with the other foot locks the rope in place.

I was summoned, one week later, to The Kirya** in Tel Aviv to continue the tests. This time I was to meet a psychologist and the Unit Commander. But, I was sick. My body had reacted severely to the extreme physical effort. I arrived at The Kirya weak from taking antibiotics and running a mild fever. I was asked to wait and since I felt ill and weak, I found a side room with a bed and lay down. Eased myself into the silence.

A few minutes passed and my time for the interview arrived, but I did not know it. The psychologist, who was female, found me lying down in the side room. She sternly summoned me to her office and started scolding me loudly, "What kind of behavior is this?" "Would you behave this way on your Kibbutz?" "Why didn't you ask to whom this bed belongs?" She continued "shooting verbal bullets" at me. I experienced looking at both of us as from off to the side and attempted, without getting emotional, to reply. I thought "This is part of the tests and I have to 'play the game.'" It reminded me of feelings during chess competitions. She makes a move and I respond. At first, I barely could breathe. Attempted to gain a few seconds. Put my thoughts in order.

While I was observing us, I slowly regained my equilibrium. I was more and more at ease and capable of replying properly. The conversation became balanced. I realized that I successfully repulsed her "offensive" and was able to communicate clearly and to the point. A smile emerged in her eyes which she attempted to hide. I felt that I had crossed this "barrier."

After this encounter, I met the Unit commander, Ehud Barak.*** Two interactions from this meeting are vivid in my mind. One was Ehud possibly asking or commenting, "Can you imagine that service

** The headquarters base of the Israeli Defense Forces (IDF).

*** Later on, IDF Chief of Staff, Labor Party leader, and Prime Minister from 2000 to 2001.

in the Unit is similar to a farmer plowing furrow after furrow, continuously, with no break? Working with a tractor in the field starting before sun rise and ending well after sun set." The picture was clear. I felt OK with such demands.

The other was his asking me if I'd hiked recently. I said that I did in the Jerusalem Mountain ridge.

"What was the track?" he asked.

I started to explain the route to the best of my memory, describing the uphill path and the mountains on both sides.

"Did you walk near settlements?"

"Yes, there were two villages on the ridge. One on the left and the other on the right."

"What were the names of these villages?"

Digging deep into my memory, I said, "I do not remember…but the one on the right… had a two-word name."

"Right, Bar Giora," he said.

Somehow, I felt that here I had also done OK.

I left The Kirya and went back to the central bus station. Caught the bus to Kibbutz Mishmar Hanegev and dozed tiredly during the ride home. On arrival, I crawled back to bed and into a deep, healing sleep with the sensation of winning in my heart.

After resting at the kibbutz, I went back to the apartment of the youth movement guides in Raannana. Muli Degani from Haifa, who worked with me in the youth movement in Herzelia, was half way through his army service in the Nahal. I told him that I was accepted into the Unit and asked him what I could do to prepare for the start of the service in a few months. Muli recommended that I run from our communal apartment, which was on the East side of Raanana, to the West side and back. I grimaced. I never was good at long distance running.

That evening, I put on my gym shoes and ran. The up and down hill going west I managed. When I started to make my way back east, my breathing became labored. My legs stiffened from step to step. My pace got slower and slower. I continued running in the deserted street at this late-night hour. Felt an unfamiliar burn in my chest. Sweating heavily. Continue to run. I can imagine that Ladany, the Israeli world champion in walking, is moving way faster than me. I continue. I arrived at the communal apartment feeling faint.

Muli saw me and asked with a smile, "How was it?"

The heavy breathing, the muscle cramps and the fire in my chest that accompanied me along the run, were slowly diminishing, but their impact lingered. I said to Muli, "I will not continue to train. If I survive, I survive. Otherwise, I will go to another assignment." I could not see myself running up and down the hills along Ahuza, Raanana's main street. Muli, generously offered to accompany me during the training, but I thanked him, dismissing the "opportunity" to experience such a run.

Muli responded, "Even if you have the capability to run the entire daily track, it is not enough. You have to run at the rate the commander is running, not lagging behind."

I said to myself, not all the recruits will arrive to the service after many months of long-distance running training. I decided, "I will cross that bridge when I get there."

And stopped training.

Before Recruitment

My heart is with the governors of Israel, willingly serve the people.
(Judges, Ch. 5, v 9)

My year of pre-army service ended and I was back to Kibbutz Mishmar Hanegev for a few weeks. Getting up early to work the fields. Morning, 4:30 a.m. Roll out of bed and walk to the dining room. First light. Drink tea with the "grown-ups." Receive instructions from the Shalhin* manager. Go to the tractor garage. Take a tractor and drive to Migda, our cultivated farm located six miles from the Kibbutz. Hitch the irrigation pipes wagon to the tractor and drive through the dusty roads to the Kibbutz passing by Tel Grar with its 2,500 years' layers of civilization.** Begin the mundane day's routine of farm work. Afternoons in the swimming pool. Meet classmates and youngsters from classes above and below me. Listen to every word about the army.

Few things penetrated from outside to the kibbutz. My attention was focused on my group classmates, Maayan and the kibbutz life.***We were seventeen girls and boys in the class. Communal

---

* Shalhin are agricultural crops irrigated with pipes, drops and sprinklers. At Mishmar Hanegev most crops require irrigation because annual precipitation is very low.

** Tel Grar is mentioned several times in the Bible as one of the important towns in ancient times. Abraham, who was scared of the king of Grar, "gave" Sarah, his wife, to him.

*** Each group in the kibbutz chooses its name according to its unique characteristics and keeps that name even after 50 years. For example, our group, Maayan, was known as a group of individualists—each member going their own way.

sleeping arrangements from birth. Learning, working, backpacking, youth movement, sleeping, eating, and playing. All together. We rarely went to Beer Sheva, the nearby city. We met other kibbutzniks our age only at summer camps, sports competitions and camping outings. I felt enclosed within a social bubble. Sealed off, yet content within its confines. Thought highly of myself. I was guided by the kibbutz's motto that "It's important to be good worker, a good student, and good sportsman."

During eighth grade after a day's work uprooting extra cotton plants, Akiva came to me and said: "You are a good worker." I felt elated and proud. The long hours of toiling in the fields and on the citrus plantation enhanced and strengthened our ability to cope with hardships and setbacks.

We spent hours at the swimming pool that was built from the German compensation funds given to my parents and other Nazi survivors' families in the kibbutz. We were attracted to the water and the"Catch Me" game. The swimming pool was our social center. I trained for years as a competitive swimmer. Fortunately, I had a wonderful teacher, Ami Livyatan, a prominent Israeli swimmer, who spent some of his Army service years in Mishmar Hanegev. Ami convinced a few of us teenagers to join his swimming classes. The challenging training with Ami gave us additional strength and stamina that enabled us to face a future of intense physical hardship.

I loved reading. Adventure books of Elister McClain and Ygal Mosenzon. Books that I devoured several times, like *The Guns of Navarone, Unknown World War II Tales, The Panpilov Men*, and the *Hasamba* series. I lost myself in the plots of these books, identifying completely with the heroes. I became the hero. Run, hide, elude, kill, assist. Serve to save our homeland. I loved reading the poems in the political newspaper, Lamerchav. Was fascinated with the rhythms and images capturing Israel's scenery and soul. I diligently read the kibbutz's weekly paper and later *Bakibbutz* magazine, edited by my father. I read each poem or article written by him.

As a child I remember the defensive trenches that dissected the kibbutz land and were a fertile playground for us children years after the 1956 Sinai War ended. During that war we were sent at sundown to the shelter with our caretakers and Shula, the kindergarten teacher. We had bunk beds with mattresses on each bed and one on the cement floor. I loved sleeping on the lower bed. That way I could talk with the child above and the child below. I felt protected within this shelter. Storytelling and playing were part of our routine before going to sleep.

During the period prior to the Six Days War, Zack, Yoram's father, told me, "It's not going to be simple. Tens of thousands will die. We have to prepare." Can a teenage boy understand such words? After conquering the Kotel—the Wailing Wall—overwhelming joy. The "Poles,"* including Abba, met at Jomski's place and tossed back "Lechaim," cup after cup. I "felt" the joy. The sensation of the verbal hugs among all of us is still with me.

I heard about the recruit hazing traditions a few nights before I began my training at Tironot Zanhanim—Parachute boot camp. One evening I found myself sitting in the room of three guys from Arava group, Yoram's class, who were about to complete their Army service. They told me how one of their commanders ordered them to lift a fallen palm tree and start writing with it in the sand as if it were a gigantic pencil. The commander dictated to them word for word and they wrote diligently. The "pencil" fell every few minutes ruining the written words and they had to correct each smudged letter.

* In Mishmar Hanegev, one of the dominant immigrant groups was from Poland. They were nicknamed "Poles."

They told me how each man had to lift a "wounded" soldier on his shoulders and run. About being woken up countless times during the night to stand at attention with full gear in five minutes. In straight rows of three. If the commander finds even a loose belt, and there is always something not up-to-standard, the entire platoon was forced to run. Run to this hill. Now to this. Back and forth. Again. Again. Eventually being allowed to go back to sleep. To be awakened again in a few minutes. Line up in triple rows. And on. As they continued to tell their stories, I felt a wave of anxiety engulf and then drown me. My eyes betrayed my rising fear and they attempted to calm me, saying, "It is not so bad," and "You get used to it." I left their hazy smoke-filled room consumed with worries and went to sleep.

It was important for me to hear about the service from Yoram, who had already spent two years in the Unit. Like me, he also completed a year of social service before the Army.

The following day I met Yoram on the path to the dining room. He said nothing about the Unit itself, but he had an important message for me. "Look Ouri," he started. "As you know, Zeevi, who was highly competent, was dismissed. Nochke was one of the shining stars. His commanders expected him to become the next Amiram Levin, and he quit during the middle of a march and left the Unit. Now, only I remain and I am OK. But you, you will determine the fate of Mishmar Hanegev. Already hard questions are being asked about Mishmar Hanegev's people. If you can stand the hardships, our name will stay intact. If you withdraw, no one from Mishmar Hanegev will ever be selected to serve in the Unit." Burdened with this knowledge I went to the Bakum, the recruitment center.

The night before. Lying alone in bed in my room at the kibbutz. The window is open and the fan is on. A light wind caresses

25

me monotonously. Toss and turn. Visualize myself waking at 5:30 a.m. to catch a ride on the kibbutz transport going north. Doze and wake. Doze and jump out of bed. Sitting. Look at the clock. Half past midnight. Lie down again. Doze. Jump up fully alert. Sit. Two o'clock. Lie down. Breathe. Take deep breaths. Unfamiliar knot in my stomach. Prepare myself to get up on time. It's hot. The fan strokes pleasantly. Back and forth. I am hot. Tense and frantic inside. It's dark outside. Can't sleep. Can't doze. Get up and go outside. Look at the trees and the lawn. It is cool and quiet. Smell the air perfumed with pines and eucalyptus trees. Dry cool air. An hour before waking to start the field work. This air is familiar and calming. Breathe again. Most of my friends were recruited a year ago. Lotan to the Airforce, Meir to the Paratroopers, Yeshai to the Armored Tanks division. I did a year of social service. Thinking: *How is it for them?* Can't imagine even one part of it. Just know that they are "there." Certain that they stand out and excel. Breathe again. Say goodbye to the desert night. Enter the room and get back in my bed. The tension dissolves. Rest. Short sleep and awakened by my inner alarm for the trip north to the Bakum.

Forty years later, I arrive at the same Bakum with Elia, my daughter, who was recruited. Impressed by the differences. Hadar, Elia and I travel together. A family on a calm morning. Park the car. Buy two hours of parking on the asphalt lot. Walk with the crowd of people through the gate. Beating drums, vuvuzela blasts, and the noise of the crowd in the waiting square. Elias' friends come to greet her. She also took an additional year for social service. Most of her friends are already in the army. Looking at the electronic board searching for her name. After an hour her name appears. We move with the crowd towards the bus that will take them a few hundred yards to where they will get through the enlistment process. Turmoil. Congested turmoil.

I arrived at the Bakum alone. Caught a ride from the kibbutz to Tel Aviv and then a bus to Tel Hashomer. At the gate, Zvika from the Unit waited. He waited for me and four of my friends; Hagai, Ben Ami, Amnon, Yehuda and myself. Each looking at the others. Watching and gauging. They seem so strong. Remind myself that I am also strong. Zvika is friendly, smiling at us. A presence with quiet, attentive energy. Walks us through the enlistment stations. It is very different than my expectations and fears. A quick simple enlistment process. No waiting in long lines. It seems at every station that soldiers were just waiting for us to arrive and treat us like kings. Probably, the rumors about the Unit, the name we are not allowed to mention, move through the Bakum like a fire in a desiccated field. From the first moments in the Army we were made to feel "chosen."

At the end of the enlistment process, a surprise. "Since there is nothing to do with you right now," said Zvika, "you are free to go. Be here tomorrow morning at 6:30 a.m. sharp. You will be transported by truck to the boot camp in Sanur."

An afternoon off, following just a few hours of being a soldier? Sounds good to me.

Boot Camp

And assigned them every one to his service and to his burden.
(Numbers Ch. 4, V 19)

Sanur. When I say "Sanur" I recall a thorny valley, September 1971. End of a steaming hot summer. High mountains surrounding us and a Sheik grave on top.* The training camp is on the mountainside. At Sanur we realized that moving from point to point in camp is accomplished only by running. Walking is for civilians. Running up and down the slope gives us plenty of opportunities to improve our physical shape. And it improves.

Our platoon is comprised of soldiers destined for the Unit and soldiers destined for Army Communications. The fitness gap allows most of the Unit soldiers to assist the Communication soldiers who were rather physically challenged. What was the point of creating such a group?

From our barracks we sprinted a short distance to the dining room entrance. Running through the center of the camp there was an asphalt road rising from the camp gate to about a third of the way up the mountain slope. A short asphalt road connected the main road to the dining room entrance.

We arranged ourselves in three rows. The Commander on duty would inspect us. His mood at the moment determined the number of times we were required to run up the hill before entering the dining

---

* The tomb of a holy Muslim person. Israel is spotted with thousands of Sheik graves, most of them located on top of hills.

room. At best, only once. Usually, three to five times. Why? Many reasons or none. As he wished. After the order "Hakshev!"—"We listen and are ready to obey," which signaled that we were in "perfect" rows of three, we anticipated the familiar command: "60 seconds, you touch the uppermost electric pole and back in three rows. Ready. GO!"

We "run for our lives," racing uphill. Our breathing becomes labored and each breath shorter. Downhill we accelerate, raising an uproar like a herd of buffalo chased off their pasture. I don't recall a single time we achieved his time limit. When the harassment was finished for that moment, we were rushed to the recruit's tables in the dining room. Separated from the commanders' tables.

One Saturday evening, we completed our sixth race up the hill without meeting the time challenge. Stood at triple rows gasping for air. Thinking about the beginning of the week before it even began.

Suddenly, the commander proclaimed, "Who does not want to run this time?"

Silence.

Our tired brains working overtime, with the little oxygen left, and suspicious of his intent. The commander moves his gaze from one soldier to the next looking for a "volunteer." On a hunch, I take a gamble. As his eyes land on me, I look firmly back at him. I "know" that he will suggest, and I will agree, to participate in the show he wants.

He plays along and says, "You will not run."

I, knowing the script, at least in my mind, say, "Take someone else, I run."

"Why bother? Stay."

"No. I am running," I respond with determination.

The commander used this exchange to motivate us and ordered us into the dining room without another run. Sometimes, rarely, gambling works.

Kitchen Duty

Arrive for kitchen duty in Sanur with Noam, from Tel Aviv. We start to scrub out huge cooking pots. Scouring pads, water and soap. The light in the kitchen is grayish and weak. The intense training of our friends outside continues. They are forced to run continually. Ceaselessly. A few hours of rest from training feels good. I know too well that it is better to go back to the training ASAP. So I will not fall behind. Still, to stop for a few hours is tempting. Even to clean pots.

"I have had it" said Noam. Small, slim boy with wise eyes. I look at him, not understanding. "What is he talking about?"

"I have had it. I see these soldiers. Our friends. And I do not want to continue the training."

"WHAT?" Great astonishment. I don't get it.

"Look. It is too hard. It is too difficult for me to run and to carry. I know that this is just the beginning. And if it is so difficult for me now, it will get even more difficult as the training progresses. I am not built for that. How many grandmothers will I have to kill so that I can go home?** To get away from this demanding training? No. There are so many other options. I thought that Matkal would suit me. No. This continuous effort is not for me. The constant running and marching. I prefer to be in the kitchen."

I still don't get it. How can he allow himself to give up? Something inside me rebels. There is something relaxing about washing pots.

The air is rank-smelling and dense, but NO effort required. One can stop at any moment. No rush. No harassments. No commands of "30 seconds go, touch, back to triple rows, GO!" Easy ease. Attracted to the relief from pressure and excruciating physical demands. The

---

** In the early stages of training, a soldier could go home only for a family event such as funeral. The phrase "to kill a grandmother" referred to having a good excuse to get a few days off.

kitchen feels like another, calmer world. But at the end of the week, I would have to return home. Just the thought of the shame I would have to endure as a "kitchen worker" does not allow me to entertain this option.

How can I stand and look at Lotan, Yeshai, and Meir? Be seen by Yoram and the others? No. The shame would be unbearable. The demands of the group I grew up with, the kibbutz community, are drilled into my brain: to go to the most demanding service. Most prestigious. In the background, the unspoken quest of my father, his family. Unspoken request or demand etched into each cell of my body. From survival in dark Times in Europe. My cells insist. No words. Yet clear demand. Demand that leaves no room for another choice. Kitchen duty is just a temporary duty. I have to train.

Waking up on the side lines—not for me. The kitchen does not fit me. From the distance of the years, I can see that high-quality friends that were with us in Sanur, like Noam and Avi Nesher, chose to leave and went on to productive careers in the arts. To the ones who continued, their choice seemed unreasonable, a sign of weakness. Each one had reasons of his own. Today, I can understand and have great respect for their choices.

Stars Outside

During the nights we trained to march, shoot, run and sometimes study. One night we got back to our beds in the long barrack early. Other soldiers fell asleep at once. I was close to the barrack's entrance. The light in the barrack was pale but light that came in from outside allowed reading. I took out *Stars Outside* from my bag under the bed, the book of poems Abba gave me, and read. The sweat, running, barking orders, all faded. I was transported to another, magical world. A few moments of deep peace. Another reality.

Forgot tiredness—for a while. Lost myself within the book. Breathed in the beauty. The bunk beds in the barrack and the atmosphere of the base receded into the background. Read and flipped the page. Continued to read. Most soldiers are already asleep.

Stillness at large whistles.
The knife's glow at the cats' eyes.
Night. So Night!
Silent heaven.
Stars cloaked. *

Suddenly, a larger-than-life figure pushes open the door of the barrack. The squad commander. He decided to check on his rookies sleeping. The first sight he sees, a soldier reading. I tense up. He recognizes me.

Right away he says, "Stay in bed."

A short pause. The world halts. Transferring me from magical escape back to mundane reality. Everything seems to move in slow motion. One can see each of hundreds of frames composing during these seconds.

Then: "What are you reading?"

Show him my dad's book and say, "*Stars Outside.*"

"I understand that we do not train you hard enough if you have energy left to read."

I feel flattered and say, "Not at all. I read only for a few moments, just to have this good taste within my sleep."

He laughs. "OK, I hope you do not read too much. Tomorrow is a harsh training day."

I say, "Good night" and read another poem. Breathe in Dad. Close the book and turn over to catch some sleep. Fall asleep as the magic dissipates while the poem's taste remains with me during that night.

* "Summer night" from *Stars Outside* by the poet Natan Alterman.

Forced March

Hurry up lest he overtake us quickly, and bring down evil upon us.
(Samuel B, Chap. 15, V. 14)

We are already three weeks into intense training at Tironut—boot camp. After three short brisk marches near Sanur, I think I can handle the speed of the next march. Based on the ideas of Elias, our direct commander, I assume that if I help a slower soldier, it will be easier for me, at least mentally, to complete the march.

The next march will be 10 miles long by the Sanur area. As we begin, Hagai and I take our places behind Beny, who is heavier than both our weights combined and consistently lags well behind any. Beny, a mountain-sized man, blocks our every attempt to close the gap with the group walking ahead of us. After a few miles, we are falling farther behind the lead group. Elias joins Hagai, B' and me to form the rearguard. Hagai and I take turns pushing every few minutes. Abruptly, we find ourselves with the squad that stopped to drink.

We barely have time to drink our share of water before the squad starts moving again. Our rearguard limps along with the heavy human anchor ahead of us. Despite the chilly night and the distance of only a few miles, our uniforms are soaked through with sweat. With each passing minute our muscles are getting tighter and tighter, our breath is becoming more labored, and our hearts are racing at unfamiliar speeds. Physically we are exhausted. Leaning forward, at a 45-degree angle to Beny, the hunk of beef blocking the road. Pushing. My head is empty. Except for an inner voice repeatedly commanding, "Forward. Forward." This voice is the only reality that exists for me.

At times, Hagai replaces me and I have a few minutes to catch my breath before resuming my 45-degree position against the "Great Wall of China." Miraculously, he continues to move, accompanied by his screaming curses, unconnected to my own battle. When I feel that I am about to collapse, the squad starts to move back towards Sanur, an infinite distance away. The road back to Sanur is mostly downhill with a few short up-hill stretches. Hagai and I return to leaning forward against our mammoth challenge, and the uninterrupted pushing stretches our muscles to the maximum.

Unstoppable continuous effort. Trudging on a comfortable dusty road. Not too many rocks to stumble over. The moon lights the track clearly. On either side of the road are cultivated brown fields or un-tilled ones with dried summer thorns. Potholes are in every foreseeable direction. The Koli direction. The Koli is the groove left behind from the tires of military Jeeps sinking in the mud after a rainy day. The tall boots we wear are well adapted for the Kolis' slopes. The weight of the gear is burdensome.

The intense effort of walking sucks out any thoughts. I enter an intimate place unknown to me. Am converging towards the center of my body. The convergence is an escape from all the pain. Detached from the pain in my limbs and muscles. Thoughts that enter my mind call me to stop. The concentration on the effort of fighting this inner voice leads me to the center, the lower belly. From this center, I observe as from a great distance all the pains that have become external. I am transformed into a multi-layered creature. The gear and the gun are farthest away. Then the sweat drenched uniform. The chafed skin. Muscles and bones. I leave my body and layers and occupy an inner core that is round and quiet.

We know that we are getting close to Sanur since our commander keeps insisting that we "move faster" and starts to run. How can we run with huge B' ahead of us? Hagai and I attempt to both push him at the same time, but the combination of moving inside the rut and the need to lean forward 45 degrees works for only two or three steps. We find ourselves again pushing for a few seconds, extremely long seconds, and spelling each other frequently. I am dizzy, my mind is in a fog. The routine of alternately strenuous pushing and blissful few seconds rest allows me to catch my breath. Then, Sisyphus pushes again. The sweat burns our eyes and we are trudging forward, straining our muscles in an effort to reach the camp gate. What is the distance to the gate? We have no idea but we know that every second, every breath, every step, brings us closer.

Suddenly, the squad changes directions.

We are moving away from the gate. Just when we were so close to finishing, the squad commander, Zmora, decided that we should grapple with the "Paratroopers' horseshoe." This formation happens when a group of soldiers changes direction 180 degrees to "re-start" the march again. The horseshoe creates a crisis extending the mission, unexpectedly, a few seconds before it's about to end.

My body loses all of its air.

Since we are at the rear, we have a few seconds rest before the squad passes us. Breathe deeply. Brace myself for the continuing effort. Console myself with "it can't be a long walk." Pray that the end of this march is near. Hagai and I resume our positions and push B' again. Begin the excruciating effort again. After a few hundred yards, another horseshoe forms. Getting close to the gate. I prepare myself for yet another horseshoe. Hagai and I continue to push despite our almost depleted energy.

Within a deep mental fog, I recognize that we have passed through the gate. The gate's light clears some of the haze. We continue to run. With painful stretched muscles, we form triple ranks on the asphalt. I am in the third row. Calm my breath. Suddenly, find myself lying on the warm asphalt. Instantly stand up and get back in line. Note to myself that I lost consciousness.

Elias commands, "Go to the showers!"

I obey quickly and reach my bed, cleansed. Before passing into the realm of sleep, I recognize that I have crossed a border. First taste of crossing borders of physical and mental capabilities.

Late Night Muster

"Each one with full gear to the Yard!"

3:00 a.m. The order reverberates. We are groggy with sleep. Push all our gear into the kitbag. Run and stand in triple rows. I am drowsy.

A few commanders are with us. The Sergeant at the center. Shouting orders. "Run to the pole and back. Again. Again."

Back at row. Panting. Yehuda, sturdy and calm, on my right. Now I'm awake. What is going on?

"Attention! At ease. Attention! At ease. Attention! At ease," the sergeant starts. "What did I find an hour ago? Somebody could not make it to the toilet? Who is it? ... How do I know? Do you know what is in there?" Pointing a few yards behind him.

Tense silence.

He roars, "A dump the size of a helmet!"

Chuckles.

"Shut up! I am going to run you all night until we know who did it. Whoever did it, step forward!"

Silence.

"Sixty seconds to the commander at the hill top and back. MOVE!"

Running with full gear and we are behind time. Breathing heavily. Back to triple rows.

"WHO DID IT?"

A small soldier whispers weakly, "I."

"WHO?"

"Gilad, Sir."

"WHO?"

"Gilad, Sir." The voice is louder.

"Gilad. You took a helmet-sized dump?"

"Yes Sir."

"Drop for thirty!"

Gilad completes the push-ups and stands up.

"OK. Run to your beds."

"Yes Sir!" we roar and race to our beds for a short sleep.

Olga—Rifles and Dunes

And I will place three arrows at my sideward, aiming to shoot at the mark.
(Samuel 1, Chap. 20, v. 20)

A few weeks after boot camp began, we arrived to Olga for weapons training. We moved from one shooting range to the next, firing at the cardboard targets. The targets were held up by wooden stakes perfect for bonfires. After shooting we cleaned the rifles. Ash and sand penetrated everywhere. The commanders checked our weapons twice a day, and no matter how much we cleaned, they always found some dirt. Each time we failed to pass the weapons inspection, we "deserved" a punishment. Penalties varied.

"Run to this hill and back. Sixty seconds. MOVE!"

"One on one, to the targets and back. Sixty seconds. MOVE!"

"All the platoon! Everyone open a stretcher, go around the guard and back. Sixty seconds. MOVE!"

Some penalties were delayed. Since our training came after the "Six Days War," Eshkol's proverb "The notebook is open and the hand is writing" was used frequently. We were continually reminded that "the notebook is open." Repayment to come. At night.

That week we got to know the night hours as well as the day hours. At 2 a.m. a clamor startled us out of sleep, "Five minutes. In triple rows. MOVE!" Robs us of our never satisfied need for sleep. In the early mist, we drag ourselves and our gear to the sandy yard. Eyes half-closed and vision still blurry from the run we have just finished before…before…actually…just finished. With familiar ceremony, the commander tells us about our daily sins, serving as both prosecutor and judge. Defense lawyer? Not at paratrooper training at Tironut. When the "judge" reads the "verdict," the "executioner" arrives. To our dismay, it is the same commander. At our level of exhaustion, we do not question or speculate. We open the stretcher and start walking up the sandy Olga hills. Going up and down. The stretcher carried on our aching shoulders.

During our previous training we carried stretchers along paved roads. The stretcher swayed in a monotonous rhythm and we got used to its waving motion. Walking at night with the stretcher among the sandy Olga hills was new to us. The air was thick with humidity and had a salty sea tang. We smelled the shrubs trampled under our boots and our sour sweat soaked through our shirts. The west sea wind picked up grains of sand that pelted our faces. The amaryllis and other bushes created sandy mounds with their roots. We would come across these mounds and stumble. As we stumbled, we would move the other foot forward to balance our body weight and avoid falling. We would hurry to catch up to the other three stretcher bearers. Each one stumbled unexpectedly. Along the smooth-packed dirt road, the stretcher would swing with a monotonous rhythm. Here the stretcher gyrated to the rhythm of wild, untamed Jazz. No steadying drumbeat. Frequently, the stumbling caused all of the stretcher bearers to fall at the same time. The twisted stretcher would go sideways and the "wounded" soldier would fall as well.

On these sandy hills we would fall with the stretcher time and time again, adding more nightmarish minutes to our march. Fewer and fewer volunteers would replace us stretcher bearers as more recruits trailed behind. We huddled silently in the middle of the night. After half an hour, which seemed eternal, we stopped to drink water.

When the break that included napping ended, we stood again looking at the stretcher. A crew of exhausted soldiers.

The commander has plenty of "fresh" ideas. He asks, "Who wants to lie on the stretcher?"

Instantly, no hesitation at all, B' says, "Me." We were not ready for this prompt reply from B', who never carried a stretcher and was the heaviest soldier in our platoon. Weighing tens of kilograms more than any other soldier.

The commander could not contain himself, chuckled and said, "Get on!"

That was not "easy." We barely managed to lift the stretcher. Started to move slowly with the enormous weight on our shoulders. We succeeded in lugging him but were angry. Very angry. Walking up and down the hills we slide, stumble and fall. The stretcher falls and the "wounded" B' drops to the ground. "Stretcher on the ground. Climb on. Stretcher up!" We lift the heavy load to our shoulders again. Only seven out of twelve of us carried the stretcher. Rotating the stretcher-bearers was happening. I found myself at the right rear corner of the stretcher. Rage rises up. My rifle, an aged FN with carrying handle, held in my left palm. This rifle bothered me continuously, each step. Eureka!—an idea comes to me. I heave the FN towards the bottom of the stretcher. A cry, "Aiiiiiiii," follows immediately. My friend who carries the left corner of the stretcher notices and does the same. "Aiiiiiiii" again. This time louder. The commander turns around and hisses at us "SHH... Night discipline!" From that moment on, the rifle stocks swing frequently towards the stretcher followed by the "Aiiiiiiii" cry. Slipping on the dunes, legs cramping, shoulders aching, the heavy weight and the heart racing during the climbs—all these are minuscule compared to the pleasure of hearing the cries of the "wounded" soldier.

From Olga we marched to Sanur. That was the longest march we had completed thus far. We walked over 30 miles. We started at night. Crossed the Israeli sandy beach strip and climbed into the hills that became higher and higher at dawn. Throne trees, oaks and Israeli shrubbery with their familiar smells replaced the sandy shrubs.

A mile before the Sanur gate we saw our platoon commander, who led the march all along, leaning against his Jeep, watching us. As we walked past him, he muttered to us "You will march the last mile without me. I have a terrible abrasion."

At that moment I felt that my world turned upside down. What about leading by example? How could the platoon commander, a living legend, stop marching? Stopping only one mile before the finish. Stop walking with raw recruits? And some are from the Unit? These questions continued to occupy my thoughts in later years.

I could not grasp how the commander who told us at the beginning of boot camp, "Some of you are from Communication and will need to work hard to complete this training. Some are from Orev Unit and will need to work much harder. And, there are soldiers from the most demanding IDF unit. And they…they…I do not know if they can survive what we have in store for them! And the training here is just the beginning—*Siftach**—for what is waiting for them after boot camp."

Leading by example becomes a guiding principle for me throughout my IDF service. This commander's example, or lack of one, becomes central to our relationships with our Crew commander.

Near the end of boot camp we are completing a stretcher march near Sanur. The ascent is demanding. We are falling behind and

* An Arab word meaning "beginning" or "opening." Usually, merchants in the market are eager to sell the first item, and they give a discount to the buyer and make a *Siftach*, a "blessed opening" for the day.

opening a larger and larger gap from the commander. Our direct commander, Elias,* takes advantage of the opportunity and stops.

"I see you are struggling." Elias looks at Amnon. He knows him. Amnon spent many hours at sea with Maagan Michael Kibbutz members. We are standing. The stretcher is resting on our shoulders. "Each one of you is barely carrying a quarter of the stretcher weight. I'll show you what you can do. Amnon, come over and carry the front half of the stretcher. Alone." I watch. Amnon positions himself below the stretcher.

"Let's go." Elias moves forward and the stretcher follows. After a few dozen yards he stops. Invites two soldiers to replace Amnon. "You see what you can do?"

A few hours before our graduation ceremony from boot camp, Elias gathers us together. He is emotional. "You are the best soldiers I have ever commanded. I truly hope that we meet again." Perplexed we replay his words in our heads. Digesting the fact that boot camp is over.

Just before we leave Sanur, Tamir, who stood out as one of the most reliable and humorous among the communication people, approached Yadi and declared, "I will serve with the Unit." Promised and made good on his promise.

---

* Elias is the direct commander, with a team of 10 to 12 soldiers. Zmora, the squad commander, has 30 to 35 soldiers. The commander who led the march from Olga is in charge of 100 soldiers.

First Dismissals

Scattered sheep of Israel, lions have driven them away.
(Jeremiah, Ch. 50, V. 17)

After boot camp we went home for a weekend vacation. We knew that the next phase of our training would be the parachute course at the Tel Nof air base. Not all of us would be there. The first dismissals would take place at the Tigers Steak House near Geha Junction. There we would be told who will advance and who will not.

We arrived early in the afternoon at the "Tigers Steak House" and sat outside. Pleasant sun sinking to the west. We compared notes about our vacations. As a young sergeant wearing red boots, no beret, approached, our alertness rose. The atmosphere was thick with tension. "Come," he muttered and moved to an area a few dozen yards east. "Sit down." We sat.

As he read the first name, we knew that all the people on the list will be dismissed. Name after name. I am still in. Tension is mounting. I am increasingly anxious. And what if my name is read as well? I silence this frightened inner voice. Quiet around us. Soft afternoon light. Storms inside us. Each new name cuts through the air like the sound of metal scraping glass…the guillotine falls.

Another name. Fear mounting. Pray.

The barley by the road sways gently in the soft breeze. Cars pass, heading west on Zabotinsky Road towards Tel Aviv. A few people walk near the Geha Junction. A full vibrant world exists beyond, but not for us. Nothing exists except the voice reading off names in front of us.

Names are read one by one.

Anxiety continues to rise, and so does my prayer. My breathing is flat. My left thigh muscle is tense. Another name. Another soldier dismissed. As more names are read, fewer of us remain. Maybe the next one is me? I fight my anxiety by going numb—denial and an effort to move to another reality where there is no "name-reader." My eyes are open but I do not see a thing. The whole world comes to a halt. I am in a mist.

Dad. What happened to him? I move through a time tunnel. See myself through his eyes. After the German soldiers passed through the Krakow Ghetto firing their weapons. Shooting for the hell of it. Whoever was in front of them. No reason. From the basement of 5 Tragova Street, he heard what was happening in the plaza adjacent to the Pankieviz Pharmacy, two houses away.

It's over.

The paper is folded up. The atmosphere changes sharply. Suffocation transformed into pure oxygen. My lungs fill with air. The feeling of being strangled is gone. The air passages open up. I take a deep breath.

"Up. On the truck," says the sergeant who used to be so intimidating.

Suddenly, everything is so simple. Tell myself that I survived. I am alive.

Look at the others who head to the other truck. Feel my belly cramp. Friends until a few seconds ago, members of our team, now separate and moving further away. Each one of us embraces a new reality and leaves the past behind.

As I go to "our" truck and climb on, the uncomfortable feeling is fading. My belly relaxes. My sense of relief is growing and transforming into pure joy. The sun continues to descend in the west. The truck moves. Tension is released. Numbness replaced by light conversation and stories about home.

Twenty-three of the thirty-two who began boot camp are left. The thirty-two were chosen from among thousands who wanted to join the Unit.

Jumping from 3,000 Feet

I returned and saw under the sun, that the race is not
for the swift, nor the battle for the heroes.

(Ecclesiastes Ch. 9, V. 11)

We began parachute training after boot camp. We were twenty-some soldiers and learned that the Unit would be accepting just one crew. We did not fully realize the implication of this information and repressed it. Acknowledging it would have required that we deal with the personal consequences of another round of dismissals at some point in the future. Instead, we continued our attitude of "business as usual" and the defense mechanism of bottling up unpleasant truths served us well, then, and in years to come.

The Unit wasted no time. To perform the winnowing process, Uri arrived. He was born in Jerusalem and had a compact, well-built body. He took it upon himself to enhance our physical performance. Since he enjoyed long distance running, we began running after the usual daily parachute training.

Our first run was to the Tel Nof Junction and back. Four miles each direction. Long distance running was my Achilles heel. It was difficult for me and I practically always arrived last. Once, late at night, we arrived back at the base after the eight-mile run. To refresh us, Uri decided we should practice carrying a "wounded" comrade. We were ten pairs of soldiers each weighing about the same. Adi and I were a pair.

In my mind, "wounded" comrade carry practice belonged back at boot camp where it was a form of harassment. Being harassed, I felt demeaned and insulted. I was frustrated and angry and did not want to make the effort. But, reality has its own demands. While feeling angry and humiliated inside, I found myself loading Adi onto my shoulders.

The instructions are simple: Each pair attempts to lead the pack.

"I will walk ahead of you and will tell you when to switch positions," says Uri. "When I order to change, the carrier becomes the 'wounded,' and you, just keep up with me. Simple."

The "pleasure" of being carried as the "wounded" is that all the internal organs get an intense deep massage. Too intense. The massage dovetails well with the walking running rhythm of the carrier. The new "wounded," who felt OK just a minute ago, now feels nausea and a strong need to vomit. These feelings are accompanied by dizziness, so every "wounded" wants to terminate this nightmare quickly. What awaits him when he becomes a carrier again is also not pleasant.

As soon as I loaded Adi on my shoulder, I felt like I was carting three times my body weight, and my legs felt like they were about to collapse. To avoid buckling I locked my knees. I could barely breathe since the weight on my shoulders was crushing my lungs. I heard Uri say "Move" and attempted to run.

Since carrying the "wounded" was not challenging enough, we performed the drill in a plowed field. The clods of earth have a known attribute. They always stand directly in the path of the carrier. So, you know that you will trip over a clod with each step. The inner monologue is: *Will I fall this time or will I be able to get my back leg underneath me fast enough to avoid the fall?* The back leg instinctively moves forward as the front leg trips on the clod. All efforts are focused on avoiding a fall. Falling means that all of the extra weight on my shoulders will be pinning my body to the ground and planting my face in the dirt. Still feeling angry and humiliated, I find myself lagging behind the other pairs by a few yards.

"Change!" We hear the order.

Adi, with boundless energy, lifts me up and begins to run. He pulls even with the other pairs and then passes them.

"Change!" The order again.

I switch with Adi and move forward but quickly fall behind again.

This ritual becomes the routine. Adi races ahead and I fall behind. Adi, justifiably, starts to get angry with me. I continue to wage an inner battle of feeling demeaned and humiliated and am thus unable to give maximum effort for the drill. At the same time I feel like a "shitty" friend who can't carry his own weigh. Fortunately, after a few more orders to "Change!" we were permitted to return to our tent for the night.

Towards the end of the parachute course, after our daily long-distance excursions, we were ordered to run from Tel Nof to Gedera. "A truck will take us back," Uri said.

The pace was faster than usual. At the Tel Nof Junction, H' decided that he had had enough. Short discussion and Uri orders two soldiers to escort H' back to the base. We continue to run. I trail behind the others. Grinding my teeth but continuing to move forward. Uri and many of my friends arrive at Gedera in good shape. I barely get there. Wholeheartedly ready to climb into the truck. We sit for a few minutes waiting for the truck. Uri drinks tea and we drink water. The truck does not arrive.

"Well, let's go back to base," says Uri.

A minute later we are running again.

Until the Tel Nof Junction I remained an integral part of the group. After the junction I began to fall farther and farther behind.

No matter how fast I run, my friends run faster. I recall Muli telling me that allowing the gap to widen means that I was broken. The gap widens but I refuse to give up. I'm unwilling to stop. Uri sends Hagai and Lony back to help me. They offer to push me.

I chase them away. "It is already hard enough and I don't want your assistance," I say.

Hagai runs back to the group and Lony stays beside me. I continue to run. I am not sure how fast I am running. Walking might be faster but I continue to run. Slowly, but still running. I can't see what is around me. Am focused on one thought. Go on. On. On. The distance between me and the rest of the group is large. The road starts to curve about one mile before the base gate. Continue to run. Mind hazy. Vision blurred. Continue.

Lony, who realizes my situation, fetches Uri who suddenly is beside me.

Through the haze I hear Uri telling me, "Stop running. Your face is completely blue."

I murmur, "No." Continue to run.

"Stop running. It is an order!"

I hear but try to ignore him.

He stands in my path and physically stops me. I surrender.

"Lony, escort him back to base. Walking!" commands Uri.

We walk back to the base. We cross through the gate, enter the tent and I collapse on the bed.

I fell asleep almost instantly. A few hours later I woke up. Crawled to the shower and then back to bed. The next day we had no training. I stayed in bed exhausted, recovering from the extreme effort. Hoping that the training in the Unit will be different. That morning, Hagai took two other soldiers with him. Just for the fun of it, they ran to Gedera and back. I slept.

Two days later the parachute course ended and we received our "wings."

Four more people were dropped from training. Sixteen of us left. We start to feel "chosen."

Basic Training—On Steroids

Everyone that lappeth of the water with his tongue,
as a dog lappeth, him shalt thou set by himself.

(Judges Ch. 7 V. 5)

Like Gideon's elite warriors in the Bible, we were also chosen.

"You see the target?"

"Attack it so you can shoot it and the enemy can't shoot you."

"OK. At my order, commence."

"For the dry run, let me check to make sure you don't have a bullet in your gun."

"OK. You are clean. Ready?"

"GO!"

I find myself jumping toward the nearby rocks and crawling the remaining distance with much effort. Breathing faster. Yelling, "Fire, fire, fire." Dash a short distance forward, leap and crawl. When I played like a "suicidal" soccer goalkeeper in fifth grade, I did not leap like that.

Kule, a Unit training area. We arrived here early morning from our tent encampment in the nearby field. We loaded our gear on the truck and traveled for a short distance to the training grounds. We secured cardboard targets with screws to wooden stakes driven into the rocky Kule soil. In some places we could not drive the stakes, so we supported the stakes with rocks that are abundant in Kule. Then we went back to the circle where most of my friends were sitting and joined them. We received instructions from our commander and started the drill, one soldier after another.

In what order should we proceed? The question hangs in the air unanswered. In the meantime, we wait. Each soldier has to go through the shooting course trail while the commander monitors him. The rest of us wait, a group of soldiers on the hill's slope, happy to rest. We talk and tell stories about home while looking west towards the blue sea shore. Below us the strip that leads to Tel Aviv and the shore. Beautiful green carpet.

From these hills and mountains our Jewish ancestors looked down at the beaches of the Philistines. Looked with awe and fear and happiness that the many hills and mountains to their east would protect in case of attack. We, young soldiers during 1972, are strong and have no fear of our surrounding enemies. The Six Day War proved our military supremacy. This feeling of superiority turned into conceit and we, the "chosen," internalize that arrogance.

An orange butterfly lands on a yellow flower. I observe. The butterfly is motionless. A sculpture. I stop breathing. See the black dots on its wings. No movement. Eternity. Flower and butterfly. The whole world stands still. Tiny tremor moves from right wing edge to the left. Again motionless. Contemplate the beauty of the orange and yellow. The other flowers surrounding us, my friends, the hills and the plateau below do not exist. Only a butterfly and a flower. Frozen in six second of eternity. I am attentive, focused, exhaling air slowly. Inhale. A light wind carries the scent of spring in the midst of the winter.

Boom! The noise of a gunshot bursts the idyll. The butterfly flies up and away. I hear the voices of my friends and am back in their world. I refocus on our training.

We sit in the circle and wait. One of us, on watch, looks beyond the circle to notify us when the commander returns after each training run to signal the next soldier forward onto the combat trail.

Yehuda returns from the trail. Sweating and panting. His face gives nothing away. He removes the army belts, drinks and sit down with us. We ask him what to expect.

"At the first target you must reach the rock on the right. Shoot from the left side and exit from the right side. Go 45 degrees to the right. At the following rock, you can rest a bit and re-load cartridges," he says.

With each explanation the training trail becomes clearer. Where one can fail. Which rock can be used as a temporary hiding place and which rock requires a quick jump. From what rock it's better to shoot from the right and dash from the left and which one require the opposite. Each one contributes to the body of knowledge by sharing his experience. There is commitment to support each other.

In a wide wadi in Kule, confined between two steep hills, we practiced throwing hand grenades. Some threw a spray grenade and some a shocking grenade. The spray grenade will explode four seconds after it is thrown while the shocking grenade will explode only if it hits the ground hard enough. The wadi bed was covered with low shrubs, wild barley, wild flowers, acacia trees and limestone rocks of various sizes scattered about. The spot chosen for throwing the grenades was behind a formation of limestone rocks. We sat a hundred yards away near a curve in the wadi, and the commander stood at the grenade post waiting for each one to take our turn.

Yadi is telling a story about a parachutist who took the safety pin out of the grenade and dropped it. The parachutist and his commander were stunned for a second and then jumped into action, tossing the grenade away from the post and lying flat on the ground, digging into the dirt. As the grenade exploded, the fragments flew over their heads.

I imagine how the grenade leaves my hand but doesn't travel far enough. No place to hide and no safe distance from the grenade. Imagine myself dropping to the ground, feet towards the grenade, and covering my eyes and balls. Shaking with fear of the pending explosion. Terrified. Can't see beyond. Continue to shudder. My legs are actually jerking.

Another team mate goes and throw a grenade. The noise of the explosion. Almost everyone has taken a turn already. The last returning thrower says, "Bring all the gear to the commander." We pack up and go out to the 'throwing post'. No spray grenades are left so those of us remaining will have to throw one of the shocking grenades. Because they are not as dangerous, all of us will watch. Hagai takes a grenade extracts the safety pin, takes a few steps and hurls it. The grenade flies dozens of yards, hits the ground and explodes.

My turn. I hear the order, "Take out the safety pin. Release the lever. Throw!"

I take the pin out, let the lever go and throw. Something seems to hold me back. Feel strange the way I throw. The grenade travels only a short distance, hits the ground and does not explode.

The commander shouts at me, "What are you. A GIRL!? You throw like a pussy. What kind of throw was that?"

I feel like a fist hit my belly. Do not let my face show that I'm suffocating. Don't know where to hide. My friends laugh at me. There is something friendly, affectionate in their teasing laughter. Even caressing. I look for something to deflect the attention from me but can't find anything. Feel the shame burning inside and attempt to suppress it. Try to think of something else. Yearn to exist in another place, another reality. To evaporate. I am angry at myself, ashamed and embarrassed. Why did it not work for me? The commander takes out his gun, approaches my grenade. After a few shots, it explodes.

"Pack the gear!"

We are moving to the next training ground. I am still embarrassed about my throw "like a Girl." The shame engulfs me every moment I am not busy packing and moving. As we are sitting in the back of the truck, the "pussy" throw bothers me less.

As I later recall this throw, I can smell the scent of the bushes and flowers, and see the rocks and the fading light before sunset. Recall the throw made by Hagai and my hesitant throw. Re-experience the laughter of my friends and the biting rebuke of the commander.

The sun goes west quickly and leaves only a few moments to get organized for the night combat maneuvers. To my surprise, the night trail is simpler than the day trail and much shorter. We walk and light small torches behind the cardboard targets. We each take our turn and are replaced rapidly by the next man. We participate in a "dry" run first. Then we receive detailed safety instructions and go "wet," with live ammunition. Complete the drill and go back to the group circle.

Sitting at night in the crew's circle is very relaxed compared to training during the daytime hours. The tension before my turn is not as strong as during the day. The darkness softens things. The Tel Aviv lights shine below us in the distance. Strong lights on both sides of the runway at Ben Gurion International Airport. A few dimmer lights from unknown Kibbutzim and villages surrounding us. As we look eastward, there are fewer lights and the darkness deepens. My eye is drawn to a very weak light. Possibly a small camp fire left behind by shepherds.

The night combat drill has ended. We pack up and load the gear on the truck and return to camp. Get organized and fall asleep. We will be woken up before dawn the next morning to complete a combat training exercise in three-man teams. Our muddy boots are near our beds so we can get them on quickly. Socks already on our feet. I put on additional layers and a sweatshirt and crawl into my sleeping bag. It is important to keep warm to get good quality sleep.

Despite the extra layers, I wake up in the middle of the night shivering. Not so much cold but dying – *dying* – to piss. Resist the impulse. It so warm in the sleeping bag. My bladder insists. Attempt to ignore it. The pressure increases. My whole existence centers on my full bladder. Wide awake. I put on my cold boots and go outside. Take a few more steps to distance myself from the tent and make sure that the wind is at my back. Spread legs and move my boot laces to the outside. Release my bladder. What a sense of relief. All of a sudden I can feel other parts of my body. Not just my bladder. Enjoy the beauty of the moon light over the silent night. Breathe in the night air. Happy that got up. Go back to the tent and return to my still warm sleeping bag. Notice that there is still time before dawn. Fall asleep. As usual, my sleep is cut short.

Recall the saying of a friend: "In Matkel you sleep very well." There are variations of this saying: "In Matkel, when you sleep, you sleep well." Or: "When eventually you sleep at Matkal, you sleep well, very well."

Stretcher March Challenge

Strengthen ye the weak hands, and make firm the tottering knees.
(Isaiah Ch. 35, V. 3)

On Thursday night we had to 'buy' the Shabat, earn our right to go home. We would 'buy' the Shabat also in Tironot by completing a Thursday night march. Basic training in the Unit required that we walk with a stretcher on the Kules' hills.

The Kules' hills have a special beauty known only to a few. In spring time, cyclamens and anemones cover the fields for several weeks. The sunsets are memorable—changing colors and lights play among moving clouds, creating a tapestry of red, pink, yellow, blue, white and all the rich variety of hues in between.

During rest periods between drills, I used to sit on a gray rock gazing westward at the sunset. Observed the rich coloring and the moving clouds. Deeply impressed by the calm beauty and the profound feeling of contentment and calm in my soul. Luckily my friends did not see me during these intimate, quiet moments while this calm sense of beauty rose inside me and engulfed me.

On Thursday evening we are sitting in the dining tent, where no sunset penetrates, eating dry bread with jam and drinking bitter IDF tea. We pour plenty of sugar into the tea but cannot get rid of the deeply ingrained bitterness. After we clear the food from the tables, built of flat wooden boards supported by a wooden frame, we sit around. We hold the messy tea cups in our dirty palms and wait. The tea has a thin crust floating on the surface. *Is it dirt or just the tea extract?* Nobody bothers to figure it out or to get rid of it. We are all apathetic while waiting for darkness to descend

Soon we will re-check our gear to ensure it will make no noise on the stretcher march. We will fill the canteens to the top, screw the plastic cover almost tight, then press gently on the plastic sides to spill a few drops until there is no air left, and tighten the plastic cover so the canteen will be completely silenced. I will shake the canteen to double-check there is no sound. We'll be ready to go, yet more time will pass before we start.

During the gap between dinner and the night march, our anxiety rises. Nobody talks and nobody knows how to prepare himself for the hours of suffering ahead.

During the week we jumped, ran, crawled and splashed through the mud. All of our clothing is soaked in mud. No piece remains dry. Our skin and clothing are saturated with a mix of sweat, dirt and rain. Now we will march to finish up this muddy week.

Within the dining room tent we all start to doze. Sleeping, although disturbed, allows us to distance ourselves from the misery ahead. All of us let our heads drop towards the plates on the table while still holding the tea cups in our hands. Swarms of flies buzz around the food remains. There is a thick layer of crusted grease and crumbs on the table accumulated during hundreds of meals eaten there. That layer blends into the khaki color of the table. We are surrounded by flies, but they don't disturb us as we lay our heads on the table, resting on our hands or stretched out arms. Our IDF shirt sleeve is the barrier between the table surface and our skin. Breathing slows and deepens as I move into the kingdom of my imagination. I do not remember the dreams that replaced my thoughts during those moments when time stood still. No notion of time passing during the wait for the inevitable. On one hand, I want to get started, while on the other, I wish this moment would last forever so that I will never have to walk. My mind is blank and my awareness is focused on the part of my arm touching the table. My head rests on my palm and at times, I feel the pressure of my thighs against the bench. The night air is dense and every few moments a weak gust of wind refreshes our breathing with mixed smells of eucalyptus, tea and sweat.

"Five minutes to go!" Our brief rest is ended.

Automatically we get up and organize our equipment. We put on our military belts and load our gear. Hagai steps outside and opens the stretcher. One of us lies down on it and the march begins.

"Stick with me" The commander's terse order.

We set out at a pace that is sometime walking and mostly running for the duration of the night. Every few minutes the stretcher carriers are replaced with others. At these moments I can rest but the rest is always too short as I take my turn as a stretcher carrier. In no time, our clothing is completely soaked with sweat. After an hour of stumbling over the Kule's rocks and running to catch up, we are ordered to take our first break.

"Each trio a canteen."

We take a few sips and pass the canteen to the next man. Each canteen is passed three times. We attempt to leave the last drops for a comrade. The canteens are empty but our thirst has not been quenched.

We get to our feet and continue to move eastward climbing up a wadi that is becoming steeper and steeper. The mud sticks to the soles of our boots. With each step walking gets harder and our boots become heavier. Our steps sink into the mud and our boots accumulate another layer of mud and more weight. At times I trip on a rock and some mud comes loose. I feel, for a few seconds, that my leg is shorter and lighter than the other leg. I anticipate that my other foot will trip on a rock, unloading some extra mud and reestablishing equilibrium between my legs. As we begin thinking that we we'll have to continue uphill forever, we get to change direction and descend along an elongated hill path towards the west.

My shoulders throb under the stretcher's weight. Our uniforms are drenched with sweat and we continue to run, breathing heavily as we take turns carrying the stretcher. We shake off the mud that continues to accumulate on our boots. We eagerly wait for the short drinking breaks that never quench our thirst. Exert much effort to "stick" with the commander unable to escape getting beaten up by rocks on our both sides. Continue to trample in the mud.

My right boot starts to "open its mouth" as the sole detaches. It gets more and more loose so, in addition to shaking off the mud, I struggle to kick the sole of the boot upwards each time I lift my leg so I won't fall. Despite the fact that we are going downhill, the effort of walking with the sole of my boot flapping loose, is becoming more difficult. When I do not carry the stretcher, I attempt to improvise a repair by looping a string I find in my army belt under the sole to secure it to my boot. For a few moments I feel relief and start to breathe freely. But the Kule's rocks have other plans. The friction from the rocks tears the shoe lace after less than 200 yards and again the sole flaps loose. By now almost half the sole is separated and kicking it up is becoming a strenuous part of my walking routine.

Another inch of the sole separates from my boot. Between stints carrying the stretcher, I manage to find a piece of cloth and secure the sole in place again. What a relief! I can walk. This time the knot holds for 300 yards. I continue to walk. The sole is separated almost to the heel. We continue to walk through the muddy wadi and I feel that I can't continue walking with my boot falling apart. All my body is twisted unnaturally and my muscles 'scream' with tension. I stop to take off the boot and secure it to my military belt using the laces. Continue to walk with a boot on my left foot and only a sock on my right foot. What a relief.

My friends notice and the march halts. The commander seeing something is wrong gets closer to me, looks and asks "Where is your shoe?" I point to the boot hanging from my belt.

"Put it on!"

"But I can't walk."

"Put it on. Now!"

End of discussion. The unspoken threat is that he will order me to climb on the stretcher. Just the thought of becoming a burden to my friends rather than carrying the stretcher, ends the argument. I put the boot on and tie the sole in place with a thicker cloth. After 300 yards the cloth wears out and I continue to limp. I make my friends promise not to talk, remove the boot and tie it again to my military belt. The crew continues marching and I can walk 'normally' again with only the sock on my foot. We resume a mostly running pace and take turns as stretcher bearers again and again. I feel tingling in my leg but it doesn't bother me as much as walking with a destroyed boot. My boot bounces on my military belt. We all are weary and our mouths are bone dry since it has been ages since we stopped for a drink.

During these hours the voices of surrender are erased and my mind can't grasp ideas such as I *would love to stop* and *I have had it*. All my mental and physical powers are focused on "sticking" to the back of the person ahead of me and keeping pace step after step. We climb on a paved road as the night fades into dawn and the march nears its end. Walking with a boot on my left foot and only a sock on my right has become routine, although gradually I have lost feeling in my right leg. An odd but reestablished routine.

We have not stopped to drink for a long time and all of us are extremely thirsty. I ask myself *What is going on? Did the commander forget?* A few hours without drinking and I recall stories about "water discipline" among paratroopers where water was restricted to just one canteen during the night and how some soldiers got dangerously dehydrated. I thought that we had learned something since that time. The stories ended with the saying: "Today there is an emphasis on frequent drinking and sufficient amount of water." So, it was said. Something inside me want to shout "I AM THIRSTY!" But I don't. Continue to walk. We move through a citrus orchard. I can smell the fragrant citrus flowers and wonder whether some fruits are left on the branches. In the Kuli, the tractor tracks that we pass, I see a water puddle. I walk at the back of the ranks and feel extremely thirsty. Stop thinking for a moment, get close to the puddle, lean forward, sweep water with my palm and drink. Another sweep and another taste. Silence the thought *Maybe there is a chemical spray in the water.* Exert maximum effort to catch up with my crew and get under the stretcher to carry it. Just before we enter the base camp, I put the boot back on my naked foot and tie the sole in place with the remnants of my sock.

At the conclusion of the march, I take off my boots, strip off my sweat-and salt drenched uniform and get under the warm shower. I sit on a plastic chair washing the march's challenges away with warm caressing water.

Friday arrives and we are getting ready for Shabbat, the day of rest, by preparing for the commander order. We clean all our military equipment and put it on a blanket spread on our beds for inspection by the commander to verify that all our gear is spotless and in order.

Then we have a crew discussion reviewing the past week and get briefing about the week to come. The commander usually reviews what went OK (very little) and what needs improvement (a lot). We barely talk during these meetings because we want to go home as soon as possible.

This week atmosphere is somber and serious as the commander talks about the "laughers" who "will not be allowed to carry on." We all think about Doron, who makes us laugh at the most difficult times. His jokes are an essential part of the crew experience that enables us to survive the strenuous drills. *Why does he pick on Doron, who brings fresh oxygen into our tension-filled lives with his jokes?* I like Doron's sense of humor so much that I told Amnon and Lony that we should have Doron living with us in the same tent so he can make us laugh.

During Friday and Saturday, my right foot swelled so badly that my mom became greatly concerned. My dad was worried as well but tried not to show it. On Sunday and Monday, my foot was so swollen that it would not fit into any shoe but by Tuesday the swelling abated and with shoes as loose as possible and nearly unlaced, I was able to walk again.

The 80-Mile March

Following the Unit's tradition: Like each crew before us we had to walk 80 miles to become a member. A month before that march, we "guarded" the border with Jorden in the Jordan Valley, "opening" the road adjacent to the fence. This practice is performed by IDF forces during the early morning, before civilians travel that road. It ensures that, in the event of an ambush, the attack will impact IDF soldiers, not civilians. Also, in case a hostile force infiltrates Israel territory during the night, their tracks will be visible on the dusty border road. If that happens, IDF special forces are called out to capture or kill the enemies before they can hurt Israeli civilians.

During other hours of the day, we trained for the long 80-mile march without knowing what we were training for. We force marched each night for 15 to 20 miles. Walk fast, stop for a few minutes for a sip of water and move on. Despite the Jordan Valley heat, the morning road opening routine felt like vacation compared to Tironot Sanur. The reserve soldiers who "opened" the border with us stared at the fence, the dusty road, and our gravel road—which ran snake-like in parallel—to look for any evidence of a disturbance signifying infiltration. We all looked to identify traces of the enemy. The dusty roads reminded me of the Northern Negev fields near Mishmar Hanegev. The sun surprised me every morning ascending over the Moav and Gilad Mountains highlighting such different landscape relative to the sunrises near my kibbutz. Also the fence interrupted the familiar kibbutz agricultural scenery. The cool morning smells and sights are all ingrained memories from a childhood growing up in the open fields under the wide skies out-side my home.

When border patrol duty in the Jordan Valley is over, home for vacation. Next week we are back. And we, without any inkling, are about to enter another world.

> *Get out of your country, and from your homeland, and from your father's house, and go unto the land that I will show you.*
>
> (Genesis Ch. 12, v. 1)

And, on we walked.

We moved from one world to another. Abraham's deserts wait-ed patiently as expected. Also, Abraham broke molds and statues. Also, Abraham had to march endless miles and uncounted scorch-ing hours through the arid heat. The world unraveled and reknitted from the same fleece.

On we walked.

Mizpe Shalem. Early morning on a cliff overlooking the Dead Sea. Starting the trek to the Unit. On that cliff we meet our crew commander. An athlete with sharp facial features. Without doubt, one who belongs to the "chosen." Tension and excitement rose.

The march commenced. After a few hundred yards, the tension and excitement abated. We walked at a fast pace towards the cen-ter of Israel. West. Northwest. The road stretched across on wide dirt trails or crushed limestone. Wide comfortable tracks. After an hour we stopped to relax our leg muscles. We drank. The Herodium Mountain loomed in the distance. Herod. *What was it the historian Yosefus Flabious said about him? Cruel, suspicious, vengeful. King of Builders.* So many structures he built that remain standing after all these years. Recall the trip to Masada with Shmaryahu Gutman, the archeology expert, and Abba who was a writer for Bakibbutz. I was

10 years old, impressed by the enthusiastic explanation, the beauty of the cliffs, water pools and palace mosaics overlooking the Dead Sea and Jordan's magnificent mountains.

During the short breaks we sat on the gravel, shook our tight muscles. Repeatedly jiggled our feet right and left and our thighs up and down. Drank. Chewed on bread and jam carried by the command-car escorting us. And on we went. We marched as a single unit. We walked uphill towards the central mountain range of Israel, whose spine runs north-south. After eight hours of muscles shaking we stopped. Saving any ounce of energy for the march.

Hours pass. We cross the mountain range. The sun sets. Proceed towards the center of Israel. Now we walk like robots. Our only thoughts concern sources of our pain. Recall the book by Eliester McClain in which he discusses a theory that one can experience pain at only one point in the body at any given moment. However, his hero experiences pain at multiple points in his body simultaneously. I too experience multiple points of pain in my body both simultaneously and continuously. The thought *Get closer to the commander* occurs to me after a while. Exert muscles and get closer.

My entire existence is marching. Within the march, only the individual step exists. Footstep. Step. Step. And step. Dirt road replaces the limestone gravel as the road is swallowed up under our soles. The pain in my shoulders is most prominent. My military belt and Uzi hang like added weights. My weapon allows my fingers to hold on to something. The swinging of the machine gun and the military belt become an integral part of the experience. My uniform is soaked through in layers of sweat. The sweat moistens the shirt, then dries. Another layer of sweat moistens the shirt and dries. Only the spots where the military belt is strapped tightly to my body remain wet. The sweat reminds me that I am losing a lot of liquid. The cells in my body are dehydrated. Dryness and numbness hug each other, create a comfortable camaraderie.

Darkness falls. The air is dry. Walking, I blend into a continuum of footsteps beyond time. A continuum that comes and goes nowhere. Just the steps exist. One, and the next one, and the next. All other thoughts exist in a blessed haze. No need to pay attention to external things. Only the steps exist. Nothing of the outside world pierces my awareness. At times a thought, a pain or an emotion penetrates my consciousness. That penetration, like a comet's trail, dissolves into the steps. My consciousness can't digest anything but the steps. The overpowering presence of the steps becomes oppressive.

Break. We collapse heavily by the roadside. Drink water from my canteen. The order—"Each three a canteen," changed during the night into "Each two a canteen." When we stop, I feel terribly thirsty. Want to drink at least a whole canteen. Despite the thirst, each of us makes sure to drink a little less than the other. Let my buddy have the last sip. There is concern for each other. *Let him drink. I will get along.* And this thought is present in each one of us. This concern for each other is part of the self-respect and pride we share and helps us coalesce into a team.

Darkness engulfs me. I close my eyes for a moment. The breeze makes me aware of my body's sweat. Drink again. The canteen is empty. Screw on the plastic cover automatically and put it back into its pouch. Make sure the canteen is packed tight. The commander gets up and starts to walk. Automatically we get up and follow him. I am positioned near the center of the walking group. My right palm caresses the pouches on the side of my military belt, ensuring all items are tightly in place. Left palm follows. Jolts the military belt. Might sit better on my shoulders or reduce the weight. The pain of restarting the march floods my senses. Each muscle screams, "I am here!"

A thought: *Don't take a break.* Then another thought: *Stop more frequently.* The penetrating chill accumulated during the break dissolves into the renewed marching pace. My muscles warm up during the first 50 or 500 steps. Who's counting? The transitions are difficult. Each time the commander starts walking after a break I think, *Let's give up, ride in the command car and start a different life.* My mind is fuzzy and can't consider this option. As soon as my teammates stand up, the thought of quitting is silenced. I get up and walk. Each one flows with the others on the trail, into the march. Into the step. We are again contained within the step. The steps become meditative. Step. Step. Step. Filter out anything that is not marching. My whole existence is the step.

How do I get inside the step? Attempt to recall the experience. It is hard to describe this physical and mental experience…

Within the step, thoughts condense from the surrounding space into my body. Thoughts transform into sensations. Pain sensations range across the colors of the rainbow, invading my body. My being is migrating to the center of my body. Enters into a protective ball below the navel region. Thoughts and pain sensations become blurred. The ball becomes the center. The pains outside the ball are fading away. Inside, it is comfortable and protected. Dissociated. Guarded. The sensation of "entering the step" occurs only after many hours of walking. I can enter the ball only after my energy is drained. As I enter the ball, I belong to another time where the movement of each leg is registered in slow motion. Each fraction moves slower. Entering the ball in the center of my body requires stripping away anything that belongs to me. I do not feel my clothes or the weight of my gear. No past. No future. The surroundings exist at a distance. Am outside time and space. Just a small ball of existence throbbing.

The tip of my back heel makes an impression in the trail's dust. The weight starts moving towards the crepe sole. More weight moves to the heel's right edge. Most of the heel is pressed into the earth. All the crepe heel's sole is crushed. The toes of my left foot thrust forward. My body's weight flows into my right foot. From heel to arch to pad. Left leg is lifted and moves forward to commence its circular movement. From within the center of the ball I flow together with the weight towards my right foot. My awareness flows in the same circle. A circle outside me. I am within the protected ball. In infinite-time. Engulfed inside the step.

My father's stories surface. Standing hours, freezing in the ranks as a prisoner at the Birkenau Nazi concentration camp and during the "death march." *How did he survive?* The thought evaporates with the next step. Back to the step.

There is a long way to go. I limp. Seeking strength. Praying for support. Shame lifts its head, pushes me forward. A stubborn thought commands: Go. Go. Go. All my muscles ache. The machine gun swings. Focus on the back in front of me. Integrate with the walking crew. There is something simple in integrating with the walking crew. Strings connect us. Pulling each of us forward into the rhythm. Into the group. Feel the comfort of belonging at the center of the walking crew. Suits me. Back to the step.

The gravel road is forgiving to our feet. Soft. Still, each step transmits its own pain message. The smell of the hyssop fills my nostrils and blends with the muscles' pains in my head. The painful step continues to be the center of existence. Walking exhausted. Do not know where the energy to keep going comes from. Completely empty. Continue to walk and merge with the step.

The night stretches before us. I am exhausted. Continue to walk with the crew. We descend from the mountains into the hills meeting the Israeli strip running north-south east of the sea shore. The darkness of the night touches the darkness of the abyss my father left behind. Europe 1939. 1940. 1941. '42. '43. '44. '45. I have to move. Each stopping will enable the abyss to devour me. Night is at its darkest point. Its nadir. The abyss nips at my heels. As long as I march, I can hold it at bay. Separated. Underneath me. Influencing. Pushing forward. Nourishing the fear. Continue to march and transfer the landing of each heel into the next step. I touch the abyss but move beyond it. The crust of the abyss is thick enough to hold me as long as I keep walking.

The morning freshness mixed with the scents of eucalyptus caresses my tormented senses. Touches but can't penetrate. Farming fields. Vegetables, plowed field and citrus orchard. Lane of eucalyptus trees. First light breaks. Misleading. Only a beginning of reality. Grows brighter. We enter the twilight region. As night is transformed into day, realities intermingle. Existence is fluid. Made and recreated. The murkiness of the night eroded by the light steadily penetrating to more and more spaces. The abyss from cold Europe loses its power.

A sense of empty body carried by two painful legs. My feet burn and hurt. Shoulders ache. Light gets brighter. At once, I can breathe.

We are getting closer. Some turmoil I can't follow. Not interested. Hear a rumor: "A bit further and we are there." A drop of happiness poured on the exhaustion, detachment and mechanical motion. The expectation ignites small glimmers of gaiety. The happiness becomes real within the painful awareness. I see a few shacks and an unfamiliar soldier who stands by a fence crossing the road. He shouts at us, "RUN!" Gather my remaining energy and run with my friends the last 90 yards towards the shacks.

Inside one of them, a sergeant shouts out orders I can't hear. Have to do 20 push-ups since I did not hear the instructions. We have to run here and there and then to stand in three rows. The sergeant orders me to drop and do more push-ups. I collapse on the asphalt. Look up with blurry eyes.

"Hey, calf. Down," he shouts at me.

I can't even feel insulted. So tired. So bleary. What does he want?

"Back to the line!"

Can see my squad mates. Tired, dirty, faint. A wave of joy moves through our ranks.

With an air of festivity, they tell us, "You have arrived at the Unit!"

We are sent to the showers.

All the turmoil at the end of the march is a blur. The shower washes off layers of sweat accumulated during the last 25 hours. Each motion slow with numbness. Grab a plastic chair and sit under the shower. Never thought that it could be such a pleasure to take a shower. I find my way to our tents. We will live in these tents for the next six months.

We all fall into the field beds lined with rubber air mattresses and cover ourselves with our open sleeping bags. Fall into a deep sleep. Wake and take a sip from the water canteen I placed near my bed before. Drink and sleep. Wake again, drink and fall asleep again.

The next day I limp towards the dinning-room. All my muscles are knotted and painful. The bone connecting the heel to the toes on my right foot, really hurts. The commander suggests that each person who feels a pain should go to Shmulik, the Unit's doctor.

Shmulik instantly creates a warm nurturing atmosphere. "You know that humans are not created for walking. It is an evolutionary mistake. I do not understand why you walk so many miles. It's not healthy. Look at the apes. They are assisted by their hands. All this idea of standing erect is a stupidity and against the laws of gravity."

I start to smile.

"Where do you feel pain?"

I show him the outside of my right foot. He touches the area. "I am not surprised. You walk too much. You probably have a bone fracture."

I look at him questioningly.

"You will need to rest for a few days. All of you will need to rest a few days since you have many other things to do. Just make sure you do not walk further than the dining room and back to the tent. And remember, it is not healthy to walk." He chuckles. "Running is probably a better option."

Both of us laugh.

Return to my tent happy and continue to rest.

What happens during the time of service? Why do people make such efforts? What propelled us? Am attempting to explain to myself: What happened to me? What happened to us? Whatever transpired seemed completely natural to us. I think that part of the processes and the experiences relate to the mindset that is created during the training of combat soldiers for the Unit. To an outsider such training might appear to border on insanity or deep inside it. What sane person will walk when his feet are torn and blistered, his muscles knotted, his shoulders bruised, his back spasming, his ankle sprained? Why should he walk obediently, fully committed to completing the mission? The ability and desire to break through any conceivable barrier are based on the infrastructure ingrained in us during our training.

To outsiders the missions might seem impossible. Within the framework of the Unit it all seems coherent and understandable. The reactions, thoughts and actions of the combat soldiers stem from the tough standards and ingrained procedures that define their existence. This is "the air" we breathe. Over the years a tradition has been created and proven effective time and time again by the extraordinary achievements of the teams in the Unit. The process seems almost scientific. Recruits arrive at the Unit competent, but no more so than people in other branches of the services. They are strong, but no more so than people in other places. Each comes with ambition and a competitive mentality. However, these traits are shared by other soldiers in the best IDF units. We were told that we can perform the impossible. We were challenged to navigate trails longer than other crews in the Unit. So we were told. Longer is relative. To walk each night 28 rather than 25 miles is harder. The "little" difference of a few more miles, of making an extra effort, is part of building a combat soldier and a high functioning combat team in the Unit.

How did we deal with this "little extra" difficulty?

With grit and determination.

Determination

Is my strength the strength of stones? Is my flesh determined as brass?
(Job, Ch. 6, Vs. 12)

There are two stories that embody for me determination, endurance and dedication to achieving a goal.

Lony is an excellent runner. During the gym practices, he was always among the first to reach the finish line in the one-mile run. Carrying weights was harder for him. His slim build made it harder for him to carry loads that were more suited to pack mules. Lony compensated for this limitation by boundless endurance.

On one of the hardest and longest exercises, Lony sprained his ankle after we landed on shore in the area of Giser-A-Zarka. The drill lasted for a full week and included marches covering dozens of miles each night. The sprain was serious and the ankle began swelling immediately. The commander ordered a stretcher and Lony climbed on the stretcher with visible dismay. We walked with the Southern Karmel hills rising on our left and started the ascent up the Taninim Wadi. We all knew that many miles lay ahead of us and that, at the end of each night, we would have to hide for the entire day.

After one mile, with no warning, Lony climbed off the stretcher and started talking with the commander. He wanted to continue walking. Both Lony and the commander were born and grew up in Jerusalem. They moved a few yards away from us and started to argue. Arguing with a commander is an unimaginable, unheard of act, no one else would have dared. The essence of the argument was another thing none of us dared to discuss. Lony insisted that he be allowed to walk. His argument: "I will walk a bit and when I can go no further, I will climb on the stretcher."

After a while, the commander agreed, probably thinking that Lony would walk one or two miles until he couldn't walk any further and then climb back on the stretcher. One mile passes. The second as well. Lony walks. The drill continues. Walking pace accelerates. Lony walks. After two-and--half miles, we stop for a drinking break. We all sit on the ground and pass canteens around. Lony undoes his boot laces and tightens them around each ankle. He adds a special knot to support the ankle. We continue to walk. Each one sinks into his personal reverie. Marching. One mile after another. Midnight comes and goes. Entering third watch. Walking. The issue of Lony persists but is ignored. Night fades into dawn. We reach a hill overlooking the Izrael Valley. Spread out in teams of three on the slope, preparing ourselves for the coming day. Each team is secure in its hiding place. First light is rising. Two sleep while the third stays awake to guard. The commander checks on Lony who repeats, "So far I am OK. When I can't, I will climb on the stretcher." Their "contract" remains intact.

As darkness starts to cover us, we pack up for the night walk. Lony dresses his injury with an elastic bandage and tightens his boot laces hard. We move into the falling darkness. Nothing is said about the pain of his swelling ankle. We all walk. Everyone of us is in a battle to survive this challenging forced march. Each exerts effort not to lag behind and keep up with the crew.

Thus, we continued to walk. Night after night. And, at the beginning of each night Lony was asked and responded, "When I can't walk, I will climb on the stretcher." His ankle continued to swell. Each of us experienced his own fair share of struggles during the dozens of miles we marched each night. Thoughts came uninvited: *That's enough. I would love to stop.* These thoughts were more frequent at the beginning of the night. Not right at the start but after 10 or 12 miles. The pain emerged from one of the muscles or joints and screams. But the scream was countered by a louder voice: *Who am I, thinking of stopping while Lony is still walking?* It chased away each thought of stopping. Night after night. The cacophony of pain would burst into my head but be chased away by the other voice.

Recall Dad. He also walked. Only three of ten completed that march. Seven did not. Seven fell. Shot. Frozen. Bitten by dogs. Beaten. Collapsed. Only three got up. And continued. Fell and got up. Fell and got up. And continued. *Who am I, his son, to dare to stop?* The thought diminishes and I return to the reality where only the step exists.

All I wanted was for Lony to stop. I wanted him to break so that I had a legitimate excuse to stop. But no. It was crystal clear it would not happen. There was no escape from the misery of walking. As each night came to its end, we all realized that we were going to complete this torturous, endless drill. Each trail we completed during these nights was marked by Lony's uncompromising determination to keep on walking.

*What is haunting Lony? The spirit of his father, who chased Nazis at the end of World War II and joined the EZEL.** No sound from Lony. Nothing to express his suffering. My pain is annihilated in the face of such uncompromising strength. The disgrace that would come from stopping does not allow a hair breadth deviation. And we walk. No need to look around. We all sense the continued step of each crew

* A Jewish paramilitary resistance organization that operated in Palestine before, during and after World War II. One of three underground movements that fought the English occupation forces before the Israeli Independence war of 1948.

member. Each step pulls us forward into the next step. The pain gets stronger. Each muscle is painfully tight. Feet are hurting. I can't bear the pain. Seal myself off from it. Feel only the movements of my crewmates near me. Center my consciousness internally, into the ball just below my navel.

Inexplicably, we are at the fifth night. It is hard to understand how each one of us made it, but even more extraordinary that Lony is still walking. Limping and walking. Fifth night. A relatively short night. We "attacked" a deserted British Police stronghold. At the conclusion of the drill, we practiced stretcher carrying for the remaining miles. And so, only a few miles from the end of that endurance march, near the field airstrip, Lony lay down on the stretcher. We all breathed a deep sigh of relief. Although we carried Lony, we felt lighter. The burden of his pain, a sensation we could only guess at, was released. That night we were all happy to carry Lony to the light airplane that had landed on the torch lit airstrip to take us back to base.

During the flight I sat near the window. The pilot flew above the long trail we had covered that week. Even from the perspective of a fast-moving bird, the trail seemed very long. The lights of the scattered villages and towns filled me with unfathomable joy. As we were sitting high above the earth enjoying the Galilee scenery, each one of us could say, "I did it."

I have never understood how Lony walked all of that long trail with a sprained ankle. Later on, we learned that his ankle was in fact fractured. Weeks after this drill Lony's ankle was still in a plaster cast until he healed and could rejoin training.

Shai Shacham

Shai, was a good friend of mine. We understood each other deeply. There was little need for words. We were bonded by our common backgrounds. Although we came from families with different histories, growing up as kibbutz boys and acting as youth guides in the city during our year of civil service created our common language. Shai was a youth guide in "Hanoar Haoved" and I in "Machanot Haolim."* He was from Kabri and I am from Mishmar Hanegev. Same kibbutz political movement, HaKibbutz Hameuhad. So many familiar and common layers.

We enjoyed arguing about any subject. Just for the fun of it. If he took a side, I would adopt the opposite and…we would argue. It is hard to describe the pleasure of arguing with such a sharp knowledgeable mind. Even though it was clear to me that I was going to lose these debates, and I usually did, I would dig in and find more and more points just to continue the argument. Sometime another crewmate would join in, and our voices rose. Shai continue to structure arguments using points from history, human behavior, philosophy, morality and current events to the enjoyment of us all.

I can see Shai standing on the asphalt square, adjacent to the porch near our rooms, and singing "Eich ze shkochav ahad levad meez?"— "How a lone star dares?" I enjoyed the challenge of singing the high-pitched tones of the Matti Caspi songs.** Attempting to sing along with no disharmony slippage.

* "Hanoar Haoved" and "Machanot Haolim" are two youth organizations that work with 12- to 18-years-old teenagers so that they will eventually choose to live in a kibbutz.

** Matti Caspi is one of Israel's top popular singer and song writers.

ALONG THE TRAIL

Take now thy son, thine only son, whom thou lovest.
(Genesis, Ch. 22, v. 2)

The death of my friend Shai is one of the painful stories. Devotion and determination were Shai's most pronounced character traits. Shai had a slight built compared to others in our crew. He walked like a goat, swiftly and forceful. His essence was energetic and vigorous. Complete another task. At the end of each navigation trail, we would lay down to rest and he would put march gear in order. Many times we allowed ourselves to relax knowing that Shai would take care of the details. During marches with the stretcher, the "threat" that Shai would need to be carried on the stretcher was ever present. He was small and light. To ensure that he did his fair share, Shai learned to "atone" for the few times he was ordered to ride on the stretcher by carrying the stretcher for longer periods. Time and again, he would carry it without taking the breaks that were customary for all the rest of us. We would spell each other resting our shoulders by walking behind the stretcher while Shai would "rest" his right shoulder by shifting the stretcher to his left.

When Shai is in charge of the crew's administration tasks, it was clear that the work would get done. His capacity to work, especially when we all wanted to sleep after a long walk, was well known to us. Sometimes, one of us would "disappear" when work needed to be done, giving birth to the expression "He went to take a dump for two hours." Shai would continue working until the job was completed.

When we arrive at the base and unload the equipment from the truck and put it in order, Shai leads. He leads by doing. Does not talk neither explains, nor argues. He acts and we join in. The work, the diligence, the singing. The leadership of Shai was quiet. From Shai, I learned the leadership of someone walking in the middle of the pack.

Creates certainty that the effort will continue until the job is done, enabling us to achieve any goal. His leadership was by example and inspired us to adopt his high standards.

Navigating long trails was difficult for everyone. Each one of us reached a breaking point. It is not easy to walk 25 or more miles at night. And another night. And another. The body yearns for the comfort of the truck waiting at the end of the navigation trail. What is easier than "forgetting" a navigation point? What's the difference? We can see the opening of the cave from 50 yards away. We can stay on the paved road 30 yards from the water hole. We can hear the frogs. We are certainly on the trail. Why should we walk all the way to the point and then all the way back to the road? Let's keep going. Not Shai. He would always go to the cave, to the water hole.

At times the following scene would unfold. We'd get close to the navigation point and sit down on the gravel road for a short rest. Shai walks to the point and back while we rest. We all drink and rest for two more minutes. Then Shai gets up and we follow. Moving on. This scenario happens repeatedly and we are unaware that we "steal" short rests while Shai completes the mission. It remains a secret shared by the navigation crew members.

Each navigation crew that was blessed with Shai's presence knew that he would be the driving force. "Yalla, Moving!" Shai will goad us and we, who wanted just one more moment of rest that stretches like the morning wake-up into a ten-minute nap, get up and start moving. In other navigation crews the "energizer" would change depending on the mood or physical shape of the members while in Shai's crew it was always him.

In Bahad 1, the officer's training school, we took part in a live ammunition drill that involved storming a fortified target. We ran while shooting and Shai got shot in the cheek. We carried him down the hill on a stretcher. Took him to Soroka Hospital. The doctor treated his facial wound. We ran back to Bahad 1 and repeated the drill. This was to ensure that we did not internalize the trauma and became afraid to fight in such a situation. Pushing us to suppress the experience. Prove it can be done without any soldier getting wounded. After we completed the exercise, we returned to Soroka Hospital to check on Shai.

The Doctor told Shai, "A half inch to left and I am not sure I could have saved you."

Shai smiled and replied, "A half inch to the right and we would not even meet." All laugh.

That's Shai.

How can it be that Shai is not with us?

> *…weep sore for him that goeth away, for he shall not return…*
> (Jeremiah , Ch. 22, V. 10)

Much was written about that cursed night on the Hermon Mountain at the beginning of November 1973 that took Shai's life. A platoon of paratroopers became trapped on the peak, the highest point in he Middle East (9,292 feet), during a storm. The weather report predicted the storm would last a full week, which meant no helicopter could rescue them. The platoon commander requested that the IDF North Command send a crew to lead the paratroopers safely down the mountain. Our crew had the most experience walking on the Hermon Mountain and the Golan Heights and was assigned the task. Our crew, along with a few additional soldiers, packed the equipment and traveled north to the Hermon. The weather was stormy. Wind gusts exceeded 60 miles an hour. The temperature dropped dozens of centigrade below zero. The crew climbed up the mountain while the storm gained force. One mile from the peak, Shai froze to death. The paratroopers came out of the cave where they were sheltered at the peak and led my crew to the cave. The next day the storm abated and a helicopter was able to land and extricate my friends.*

I was not there. I heard the story many times from my friends. I attempt to visualize Shai's determination climbing the trail up the Hermon. Walk and fall. Fall, get up and continue walking. What no one could ever imagine is that the price of his determination could be so high.

His life.

The helicopter took my friends north. Yadi and I stayed behind. We envied them. At the lower cable station, they climbed into an armor car that took them to the base camp. The wind howled.

"The hail pelted us," said Adi. "Our efforts to connect all of us to the climbing rope seemed to take forever."

"Last check before moving out. Instructions passed from mouth to ear and the wind snatches the words away. Arranged in a row. Lony and the commander leading. After a challenging march, we pass near a deserted Russian machine gun. An ominous scene. The wind slams the hail into my left cheek."

Ascending.

"The gray rugged rocks leave a narrow walking path."

Ascending.

"The winds build. The hail, like knives, pelts my left side. Descending from the ridge top eastward. Resting. Conferring with the commanding officer. Moving on. The walking path is exposed to the

* Information about the events leading to Shai's death can be found in a report by Roni, Yonat's daughter, Shai's sister, in Ido Netanyahu's book *Hamazilim*, a chapter in the book *Zevet Itamar* by Avner Shor, and newspaper articles from that period.

wind and hail. Ascending with effort. Dark. Thoughts and feelings are dimmed. Shai stumbles. Continues. Stumbles again. Walks. The storm gets stronger. Shai falls. The pack halts. Gets up. Moves on. Over ten hours in the freezing cold. Ovad walks near Shai. Shai falls again and again. Gets up and insists to continue. Two soldiers support Shai. The rocks prohibit walking side-by-side. Shai is dazed. Barely moving. Falls again. Stopping. Looking for a shelter. The paramedic unloads his medical bag and it falls downhill. Disappears into the foggy chasm. No medical equipment for the crew. Attempts to warm Shai. The aluminum foil blanket is torn. The small torch refuses to light. His pulse fades. Ovad and the paramedic lean over Shai."

Days later, the Unit doctor told us that during hypothermia the body gives up on its external limbs concentrating its energy and blood flow on the essential organs—brain, heart and lungs.

"Shai skips a breath. Stops breathing altogether. Ovad resuscitates him frantically. The paramedic joins in. Others take turns. Ovad resuscitates him again. Massages his heart through the chest. All are dazed. Alerting the paratroopers to come out. Ben Ami moves from one to the other and scolds, 'Move! Up! Move!'

The paratroopers climb down one mile and help the crew members get to the shelter of the cave. Feelings of failure and utter exhaustion."

The mountain that became the Golani curse during the Yom Kippur War, challenged us many nights, before and after the war. Many times we completed our missions exhausted to our bones, but always we won. Not this time.

Every time we talk about that night, the sense of loss resurfaces while we try to understand what happened there. We all experienced endless, effortful struggles. At some point, the struggle gives way to a mechanical walking motion. And Shai walked. And when he fell, he got up and continued to walk. Walked in probably the same way each of us would have done. But Shai did it with extra determination.

When I imagine the scene of Shai walking in a daze, ascending the mountain, I feel distress. Tears fill my eyes and my throat tightens. Recall experiences of walking in a daze. Recall the determination of Shai and the respect we all had for his determination. And then the thoughts surface: *What would have happened if? What would have happened if there was a decision to stop when it was still possible? What would have happened if there was a decision to go back? To ask for help?*

And knowing intimately the "Unit spirit" and the determination to complete operations, I know that these questions are merely theoretical and could not lead to discussions of any practical alternative. Because they had no place in an operational crew in the Unit. Certainly not in our crew that was forged at extremely high temperatures to accomplish "unattainable" goals. Determined.

And the determination that enabled the achievement of extremely difficult tasks is the same determination that does not allow thoughts of going back or giving up. This mold was cast from years of grueling experience, walking for endless miles, both counted and uncounted. Walking for miles that ingrained in each muscle and cell and embedded in each layer of our personality. Each statement uttered today stems from understanding after the fact. Irrelevant to real-time occurrences. Completely theoretical. Whoever experienced the training of an operational crew knows that mold is not easily broken. That determination was burnt into us with a branding iron to create the Unit's core operating principle. We did not know any other way to think. And within this ethos of the Unit we lived and operated and achieved the impossible.

And Shai died.

And the burden is heavy.

Today I can say, to my dismay, and with laser sharp understanding, that we had no other option. It was our way of being. Was Shai's way of being. Unusually determined. Determined with no reservations.

And, I would dare to say...aching—**determined unto death**.

Other crew members in that accursed night missed their death by a hair's breadth. Within this existence, that frame of being, there was no room for "What if" questions. There was no room for anything except the determination to complete the mission.

We appreciated Shai for his determination. Loved him. Were glad he was our friend and that we were his friends. His presence was a source of inspiration to us all. Still is. Inspiration that remains across the passing days and years.

Today, all that left during the mundaneness of daily existence are the annual remembrance days and longing.

I miss you Shai. Your thoughts, words, smiles, your being.

I stop. Breathe.

I think that each personal story—the sacrifices made to complete each mission and the Unit demands for ceaseless, Herculean efforts—expands physical and mental capabilities. Each time we broke an operational record, the Unit demanded, systematically, that we achieve higher, more challenging goals. Again. And again. This is a critical condition enabling the building and refining of the operational capabilities of combat soldiers in the Unit.

Raising the Bar

Thou hast enlarged my steps, and my feet will not stumble.
(Samuel B, Ch. 22, V. 37)

The process of raising performance capabilities and expectations starts on recruitment day and continues through obligatory service and well into reserve duty. Expanding operational capabilities is done through continuous mental and physical challenges. We began with long distance walks and runs and graduated to stretcher-carrying marches and traversed long navigation trails. After basic training we went further, carrying weights that became progressively heavier along longer and longer paths. Moments of reaching a breaking point occur in different forms along the trails. As the walk becomes longer and harder, breaking points intensify. Each walk becomes an opportunity to further extend and strengthen our capabilities.

For a "Northern crew"* operating in the late 1960s and early '70s in the areas of the Golan Heights and Mount Hermon, improving physical capabilities was a necessity. We practiced mostly in the mountains, hills and plains close to Syria. In the north, establishing the march rhythm was not simple. We walked on a basalt rock surface carrying heavy weights. We learned to leap from point to point in fields seeded with endless lava rocks that threatened to sprain our ankles. And did.

* A crew that trains in the Hermon Mountain and Golan Heights and operates mostly in Syria. The extreme weather and terrain conditions require exceptional physical fitness and endurance on the part of the combat soldiers.

On we walked.

We covered numerous miles in the Basalt plains. The rocks' colors range from red to gray and black. Our drills were mostly during the night. At night, the basalt rock becomes dark, sucking each bit of light. I can see a dark surface. I can't distinguish the basalt rock with my eyes. I can sense the rocks through my shoes. It is imperative to adjust foot placement to the curvature of the basalt rock. When I misjudge, which happens frequently, I struggle to avoid falling. The machine gun takes on an independent life and hits the rock with a screeching sound that violates all the commando combat rules of keeping silent that we learned. I mutter a curse and continue to walk.

So many basalt rocks deflect my steps that I will fall if I don't lift my back leg swiftly enough. If I hit a rock with my back leg, I lurch forward at a run. Well, running is simple. But, running while carrying a heavy load on my back that barely allow me to walk, is challenging. When I hit a rock and start running, I pray that I will not have to run more than a few steps. Running with many kilograms of weight on my back is madness. My pulse races at extreme rates.

During these moments, I thank Izik Levy, our gymnastic trainer. He used to chase us from end to the other of the British hanger at the base. Up and down. Again and again. These races and other exercises to raise the pulse rate give us the ability to accelerate our clumsy walk into a short sprint.

On the walking trails along the Golan Heights, we negotiated on the goat paths used to move livestock, adjacent to agricultural terraces. The trails are narrow, but for us they were "restful," significantly reducing the chances we would need to sprint. For us, these paths were more comfortable than European Autobahns. Not that I knew what a European Autobahn was at the time, but the comparison turns out to be accurate.

Before each walk, I studied the night trail and counted the miles on the basalt field and on the goat paths. I would tell myself, "Here I have to be physically ready and here I get a chance to rest.

Rest? Such a "rest" involves walking up or down the path with dozens of kilograms of weight on my back. While walking I develop a dark sense of humor to keep me going, acknowledging the fact that I can't move millions of Golan rocks and will hit these stones each few yards.

We walked five nights a week. Each night had its distinct characteristics. The first night was mentally difficult since a few days have passed since the last walk. The second night was more difficult physically because the walk was longer, but I was already in the mindset of "doing." The third night was difficult because fatigue is cumulative. With the fourth night we got closer to the fifth night, the last for the week. The fifth night was the longest of all, but we already could 'smell' the upcoming weekend and home.

We start the fifth night with considerable vigor. Complete the drill and go. My back, limbs and feet have accumulated an entire week's worth of pains and exhaustion, but my heart is full with anticipation of the weekend's joy. Waiting the end of the fifth night of hard walking allows the winning sensation that comes at the end of the exhausting trail to touch me. But the taste of nearing the end is not fully satisfying. The fear that I might not be able to complete the difficult trail presents itself as an equally familiar taste. The flavors of anxiety and the desire to succeed blend. The anxiety is clear and familiar. The possibility of winning wavers. Not in the bag. Very close but…maybe not. The anxiety turns to irritation. Fear of possible failure gnaws at me. Familiar tightness in my belly. As the anxiety surfaces, I suppress it and it raises its head again.

I know that strengthening performance capabilities requires the willingness to let myself "crash" through the familiar barriers of the merely possible. I know that I am about to hit a wall. Smash into it. All my beliefs and concepts of myself will be shattered. On I go. Crushed again. And again. Collect myself, my fractured beliefs, and move on. From within my shattered psyche, I can build the possibility of continuing to walk even after having crossed my known boundaries. As I cross them, my performance capabilities expand.

In a few moments we will start to walk. Breathe. Enter inside myself. With the breathing and the internal centering, I move towards the "ceasing to feel zone." Minimal external interaction. Test my gear. Uzi, cartridges, military belt, backpack. All is properly fastened. Boots? Check the tightness of the laces and the elastic bandages on my ankles and socks. Bandaids protect sharp edges of my feet so they will not rub against the boot leather. Check again the tightness of the elastic bandages to ensure that my ankle is firmly supported. The tighter, the better the ankle support, the better my chances are of completing the night march without suffering a sprain that will snatch this night's victory of completing the trail from me.

Continue to center internally. Visualize the trail and mark in my head points of reference along the long path. Points where I can report to myself: OK, this portion is behind us. Later, I likely will not remember these reference points, but for now visualizing these points calms me. That's it. Now we get ready as a crew. Re-checking the gear and equipment. Ovad comes outside, unzipping his pants. We join him on either side, Lony and Ben-Ami mutter, "All-in-one-line." Chuckling. We all pee and empty our bladders. Light is diminishing. Drink a bit of water. Take my spot in the walking ranks. Darkness falls. We walk.

Thoughts surface: *Am dying to stop. Enough. Let something happen to me. I wish I would sprain my ankle. I am fed up.* I am aching. I feel so many pains in so many muscles. Thoughts attack me. Penetrate like arrows to my inner center that urges: Yalla,* *Go forward. A bit more.* Even though I know there are many more miles to walk. Commands: Go. Go. Go! The positive thoughts nest in my energizing center, absorbing the demands to give up and stop. My body has learned through experience that I can survive these moments. That I can overcome. Add to that the knowledge that as I complete this task now, others have completed it before me. The thought rises: *I am at least as good as they were.* And the haughty voice presses forward claiming: *I am better than they were.*

I heard from Yoram about his crew member, Uri Korn, who walked carrying a weight of dozens of kilograms on his back across the Golan Heights and Mount Hermon, for endless miles, and continued to walk when most of the other crew members were on the verge of breaking down. As soldiers in a 'Northern crew' we attempted to match these standards and exceed them. And we did. Thus we walked.

Walking throughout the dark night. Walking. The darkness is dense. With no hint of warning, I realize night is over. Dawn emerges hesitantly through the fog and clouds, and we reach a half destroyed shack that still has a roof. We stretch out on dry IDF rubber mattresses. We take off our damp cloths wet from rain, fog and mostly sweat and hang them to dry near the oven. Pull off our boots and socks and place them near the oven, too. Put on dry clothing and crawl exhausted into our sleeping bags for three to five hours of dreamless sleep. The victory of that night will be tested by the

* "Move it!"—Israeli slang originally from Arabic used to encourage someone to do something.

following night. Yet the joy of this night's achievement, the ability to overcome challenges with determination, follows me during a good, restful sleep. The struggle between thoughts of keeping going and quitting repeats itself night after night, week after week, and month after month. After years, the opposing viewpoints of the internal struggle become familiar territory and the internal debate diminishes. I continue to grow accustomed to more and more difficult efforts. This training regimen strengthens our capabilities. The endless marches carrying extremely heavy weights on our backs build our endurance and enable us to break through barriers we thought were unattainable before. As we reach the new goal, a further more 'unattainable' goal, challenges us. And then to the next barrier and the one beyond it. And beyond. Another border crossed.

And Dad escorting.

Recall thoughts that surfaced, usually towards the end of the night or just as the walk was completed. Calm thoughts but strongly present: *How did Dad walk the death march? He barely had water to drink and food to eat. How did he make it?* The desire to compare my experiences to his is tempting. Does not matter how difficult or strenuous my effort is. His experiences were much harder. There is no real possibility for comparison. Yet something in me is drawn to measuring myself against Dad, although I know that I can only infer and can't ever, really compare.

It might be more accurate to say that it is an attempt to relate. A desire to experience a little taste of what he experienced. I think that the physical challenges I endured enabled me to understand a small portion of his hardships. I understand through doing and the experience of long exhausting marches. Through my body. The effort allows me to feel close to Dad. To connect to the taste of his experience.

On the death march to Gliwice, after a frozen sleepless night, he was barely able to stand up. He told me that he could hardly move his legs, and it was only with the assistance of friends who supported him that he was able to continue the march.

In our crew we had a few tricks to encourage ourselves. We built a framework to break overwhelming challenges into smaller digestible pieces. We would say, "Within 20 hours, our situation is going to be much better." Later it became, "Within forty hours…" or more. Just the knowledge that there is a known end point leading us to a hot shower and the end of the effort, calmed us.

Humor also helped us survive. Luckily, Doron made us laugh frequently. We all enjoyed laughing which provided a great relief but the commander did not like it. In his eyes, laughing was frivolous. During the first months of training, the commander told us that we are not serious enough. As if anyone who laughs, can't perform serious work. Only today can I utter that idea with ease. Yet, as we progressed through our training and became an operational crew, laughing became extremely important and ultimately acceptable.

Our team building continues. Week after week, a Northern crew continues to build and expand the capabilities that accompany each man along the trail, through the darkness of each night, on each mission of our IDF service.

Rising determination and perseverance are not just my personal experience. Reaching each breaking point and moving beyond it takes place within the intimate framework of the crew. Everyone sees and is seen by the others. Each eats, sleeps, runs, carries stretchers, marches, navigates, snores, returns from home, pees, argues, defecates. Everyone sees me. I see everyone. Each crew member understands and is deeply familiar with the boundaries, capabilities and reactions of every other crew member. Each one of us experiences, with no filters, what happens to every crew member when he reaches his physical, mental and emotional breaking point. At the edge of human capabilities, each one operates and sees his friends operate with no mask. There is no shame nor pride. Just performing or not performing. We all stand naked. Skinless. Our nerves exposed while making an extreme effort. We are all intensely focused on performing our assigned tasks as a crew. The exposure during these extreme exertions forges an intimate and unbreakable connection among us.

Today, each human connection I know is woven through talking. Not in a crew. The bonds are forged by performing tasks together. Facing challenges and making a group effort to overcome them. Harmonious activities, walking and working deepens understanding. Discussions take place when performance is being evaluated to learn and improve. We rarely talk about 'nonessential' items. I have a hard time recalling "normal" conversations. This is part of the infrastructure established to coalesce us into a crew.

Biking to Eilat—2

Goral junction allows us to detour around the city of Be'er Sheva on the northeast side. We reach the junction and get off our bicycles. Find a spot at the side of the road, lay the bicycles down and drink water. Lie at the edge of the asphalt. Ruvi, my friend from Shoham, arrives with the pickup truck and hands us delicious sandwiches made by Hadar. The first bite explodes in the mouth with rich flavors of Tsfatit cheese, tomato and basil. We indulge and eat more. Drink more water. The other riders arrive. Rest a bit more and drink.

"Let's go" says Lony.

Yadi and I take a position behind Lony to support him during the long ascent. Chatting to shorten the uphill climb. Towards the end of the climb, Lony presses us to increase our pace and meet up with the others at the next stop. After a short argument, we agree and rush downhill. Complete the Be'er Sheva detour and start the 20-mile ascent towards Dimona. Strong wind gusts pummel our faces. Each turn of the pedals requires much effort. We settle into the pedaling effort against the wind. Pass many Bedouin shacks on both sides. I hardly notice them because my intense efforts to maintain speed and stay at the edge of the asphalt. I notice road signs, road embankment gravel, mile markers, hills devoid of vegetation and the steaming asphalt. Continue to pedal uphill against the battering wind towards Dimona. The heat intensifies and so does the wind. Drink water as we ride. My sweat dries and turns to salt. Pedaling against the wind is exhausting. Pedal, breathe dusty dryness, and pedal.

Why did we embark on this bicycle trip?

The idea popped into my head after I had a hernia operation. I thought it was time to do something physically challenging before my body falls apart. I talked with my crew buddies and we began training six months prior to the trip. It is a powerful, yet inexplicable attraction to pedal with my Unit's crew. The same crew that experienced and overcame extreme physical challenges together. To re-experience the pre-mission anxieties, overcome fears by doing the job together and experience the sense of victory after accomplishing the task and saying to myself, I did it! The feeling of accomplishment: We did it together!

And it is another opportunity to commemorate Dani Senesh and Shai Shaham.

We are getting close to Dimona. The Eucalyptus trees are plentiful. We pass the traffic light at the junction and rest in the shade of one of them. The pickup truck with our provisions arrives and we enjoy the rich taste of a sandwich and drink water. We eat granola bars and sesame sticks while we rest. Climb back on our bicycles heading towards Yamin Plateau and the decent to the Arrava valley. I'm looking forward to the steep and long descent into the Arrava in about 12 miles.

Somewhere along the cliff, Lot oversaw the smoking ruins of Sodom. In a nearby cave, where they sought shelter, Moav was impregnated, after intercourse with her drunken father. King David came out of Moaviat Ruth. I start laughing.

The descent towards the Syrian African Fault is exhilarating, coursing through my body. I let it "take" me. Bent forward over the handle bars and eyes on the road. The speedometer shows 50 mph. As the curve approaches, my fear rises. I straighten my back and slow down. Loud laughter. Extended enjoyment. The descent is less steep and turns into an ascent, followed by another steep descent down the fault. The wind whistles in my ears. Relax my leg muscles by lowering my heels. Inhale and exhale. Freedom. Let it continue. At the Arava Junction, turn right and stop at the gas station. Yadi arrives and Rovi right behind him with the pickup. Sandwiches and water. The heat is comforting. We devour schnitzel sandwiches and drink and drink. Looking at each other and Yadi and I realize that both of us have huge smiles plastered on our faces. The thrill of the descent.

One-Mile Run

It was joyful to go back to the Unit base on Sunday. Everyone brought some food with them so we could savor the taste of home for a few more hours with schnitzels and cakes. I loved the fresh flavor and smells of the yeast cake Mom baked for me and my friends on Shabat. Even today I search for that taste and smell in cakes around the world. Search and have not yet found the exact sweetness, crispiness, poppy seeds, package paper and aroma.

As I arrive I place the ironed uniforms, washed in the kibbutz's laundry, in the cupboard so smells of home and IDF are mixed together. In a few hours the IDF existence will assert itself and just a hint of home smells and memories will remain. We start to prepare this week's training gear from lengthy lists for each soldier and the entire crew, and place the equipment on the beds and sheets to be inspected and all the missing pieces assembled. My personal gun is an Uzi, so I check that in two magazines each second bullet is a tracer bullet while in the other magazines all ammunition is clean, regular 9-mm rounds and that the magazines are spotless. I rinse out the canteens several times and fill each with fresh water. My gun is clean and ready. Each pocket in my army belt is filled with combat equipment except for one that will be filled with bittersweet chocolate and dry figs that mom purchased for me in the kibbutz's markolit.*

In our room, I would slice the cake and my friends and I would wolf down the cake in seconds. Doron touches my back and measures

---

* A grocery store with limited items, sufficient for the kibbutz's "Spartan" inhabitants.

97

it with the span of his thumb and pinkie. Lony questions, "A third and two thirds?" Since my legs are shorter relative to my torso, and Doron declares, "Wahad Jab"—"Quite a back," and all of us laugh. Doron and Lony tell stories about their encounters with girls and I listen quietly, embarrassed that I have not had such experiences.

I would set aside a small sliver of cake for later, in the late night hours after we came back from the night drills. Then I would eat that small yeast cake and feel blissful.

We go to the Unit's dining room for lunch. Dany joins us but barely eats since he is still full from his mom's schnitzels. In this way, Dany 'fights' to hold on to his sense of home into Sunday and, if possible, Monday as well, feeding on the wonderful delights cooked by his Hungarian mom, Leah.

The rest of us sit and fill our stomachs with the "Nathan Delights," our name for the basic, familiar food - soup, meat, salad and bread—that the Unit's longtime "chef" cooks up. We return to our rooms sated and continue to organize the drill equipment.

By this time I am ready for a *Schlafstunde***—an afternoon nap— but in a few moments we will begin our gym drill. We complain, as usual, "Why is gym scheduled so close to lunch?" We put on our gym clothes and Izik, our gym instructor, is already waiting for us. A short warm up and then we proceed immediately to the one-mile race.

Each Sunday, Adi and I got a lesson in modesty and coping with frustration. The commander and some of my friends were excellent runners and they competed among themselves for who would cross the finish line first. Most of the crew was running in the center of the pack. Adi and I would trail far behind. Within the competitive framework of the crew, comparing ourselves to one another in each

** Originally German meaning "sleeping hour," it is also Hebrew slang for "siesta."

98

task, the one-mile run was an insurmountable barrier for me. I felt that I ran into a wall time and time again only to bounce off and land on my butt.

No matter how much effort we exerted, Adi and I were always last. We started the run at full speed and energy and exerted tremendous effort. Our legs hurt, our hearts raced and our breathing was strained. But no matter how hard I tried, I always finished last. Always last. Struggling for breath and exhausted. The frustration was unbearable. I envied Dany, Ovad and Hagai who always arrived first, seemingly without effort. As a boy, I dreamt to be like the long-distance runner Emil Zátopek, whose nickname was the "Czech locomotive." I was thrilled by his victories in the 5,000 and 10,000 meters at the Olympic Games. Long-distance running was always difficult for me, yet I wanted to be, even a little, like Zátopek.

As the gym training continued, we climbed up ropes and sprinted up and down the hangers. I was better at these activities. I managed to excel at rope climbing, and Adi and I were usually among the first in the sprints. These small victories made us feel wonderful. Yet every Sunday, I competed in the one-mile run and I knew that no matter how hard I tried, I would finish last. All I wanted to do was finish in the middle of the pack, yet I failed every time.

Galilee Navigation

And a trail shall be there, and a way, and it shall be called the way of holiness.

(Isaiah, Ch. 35, v. 8)

Navigation is a central building block of combat training in the Unit. Navigation drills incorporate all the skills a soldier needs to be a contributing member of an operational crew. Navigation is built on learning, performing, and inquiring; and more learning, performing, and inquiring. The crew members develop the qualities of precision, perseverance, determination, endurance and cooperation by completing the required tasks and demands of the mission under time pressure. Navigation drills foster the necessary culture and language that enable a crew to operate as a united fist down the road.

Navigation drills begin in the Galilee region where the mountain ranges and wadis are simple and stretch along clear directions. On the Galilee's western side, the valleys slope towards the Mediterranean Sea. On the eastern side, the streams flow towards the Sea of Galilee and the Syrian-African Fault.

We start during the day time since the light allows us to see each reference point clearly. We can recognize the mountains, hills and wadis as well as roads and villages. Only later we will be challenged to navigate at night in the darkness.

During the first day of our navigation drill in the Galilee, Yoni Netanyahu, our commander's commander, responsible for all training in the Unit, joined us. At first appearance he didn't look at all like a combat soldier. He was chubby. In a few moments, however, we felt his authority. As we reached the top of a hill near Kibbutz Kabri and gathered together around him, Yoni pointed towards the rising sun and asked, "In what direction is the sun now?"

With no hesitation I replied, "East."

"Are you sure?"

Instantly, I recalled how my father would train me in math by asking me to solve puzzles using my head only. He would fire questions at me and I would respond. The level of difficulty would rise from exercise to exercise, and I would respond faster and faster, attempting to keep up with his pace. At some point, Dad would ask me something more challenging such as "What is 16 time 19" and I would reply, "304." At that moment, he would nail me without mercy: "Are you sure?"

My brain would shrink to the size of a small grain. My confidence from previous correct answers would evaporate. I was left with nagging doubt and began to stutter, "Aa…ah…aa…," and then fell silent.

With Yoni's question I traveled back in time for a decade and was silent. Yoni took advantage of the dramatic impact and said, "Take out your compass and look at the sun, but look just for a second."

To my surprise, the compass showed 100 degrees rather than 90 towards the east. Yoni explained: "Israel is at a latitude of 33 degrees or more, and during the winter time, the sun moves in a tighter arc than in summer time. As a result, during winter time the sun rises a bit south relative to the east and sets a bit south relative to the west. During summer time it rises a bit north relative to the east." This lesson became ingrained in my head.

During that week, we climbed many of the Galilee mountains and walked down its wadis. We became familiar with the Galilee's shrubs, its mountains' steepness, the tangles of raspberry bushes near the small brooks, the olive and eucalyptus smells, and the croaking of the frogs during the evening near the springs.

We walked in teams of three, racing the other teams to the finish after all the navigation points along the trail had been identified.

We learned that the basis of effective navigation is direction and distance. One has to be able to answer two questions: In what direction am I heading now? And: How long should I keep going in this direction? By studying direction and distance, we came to know the mountains, hills and valleys. When the navigation involves descending down a wadi, there is no chance for error since the wadi "collects" us like rain drops. Navigating up a wadi is more challenging and deciding which branch of the creek to choose is most difficult. At each fork of the wadi we must choose the correct branch. Shall I take the right or the left one? A saddle is a good identification point since there are two hills, one on either side, and two wadis—one we climbed and one we descended. We learned how to identify navigation points and how to memorize the surrounding physical details. We always looked for the northeast corner of each navigation point. Afterward, we had to be able to describe that corner in detail. If we reached a ruin, we had to describe the way the stones were stacked in the northeast corner. Same for a well or a water hole. The northeast corner became a code for us and was incorporated in our "secret" culture. No matter where we planned to meet, in Tel Aviv or Jerusalem, in a building or a plaza, we all would automatically show up at the northeast corner. There was never any need for guessing where to assemble. No further coordination was required. Everyone waited at the northeast corner even decades after we had all retired from the Unit.

At the Galilee, I participated in our ninth-grade, annual school trip. We were divided into groups of four. I walked with Eitan, Meir and Zvia. We were supposed to arrive to Kibbutz Idmit, but at some point we became lost. We looked at the map and realized that we were not far from the border with Lebanon. We had no idea where we were and wandered a few hundred yards in each direction. We were scared. Are we getting too close to the border? Did we cross the border? We looked around, frightened. After a long hour, we saw a man walking on the slope of the wadi. It was our guide, Eitan shouted, "Angels! Angels!" We joined in and waved our hats. After what felt like an interminable time—possibly two minutes—our guide waved back. We were saved. Each time I am in that area, I remember hearing Eitan's shout.

On the morning of the third day of the navigation drill, we ate bread and IDF jam, an omelet that was burnt and runny at the same time, and white cheese. I drank the thick IDF tea with a drop of lemon and a teaspoon of sugar. I made a sandwich and put it in an empty pouch on my military belt. We began walking. Navigating during the day was fairly simple. The mountain ranges were clear and visible, extending from east to west. We covered one mile after another. We stopped for a short break and my stomach began rumbling. Another mile. We felt an urge to move faster, to quicken our pace. We knew that the faster we walked, the less time we would spend on our feet so our aches and pains would be minimized. Our muscles are tight and I attempt to move faster along the trail to reduce the hours of physical suffering and have more hours of blissful resting. The capacity to move faster and maintain a brisk pace for myself and my friends along the lengthy trails is heavily dependent on our physical shape. We all are focused on marching briskly. We attempt to identify the navigation point as quickly as we can and to be accurate so we can stick to the trail and waste as little energy and time as possible. My stomach rumbles and the pressure in my bowels builds. I stop and tighten my buttocks. The pressure eases yet I know that in a few moments it will return. I need to defecate. The pressure does not allow me to continue. I tell my friends, "I have to shit." I see two sizable stones standing side by side. Take off the machine gun and military belt and toss them nearby. Drop my pants and underwear, stumble towards the stones and crouch. Put the boots and pants in front of me as far as possible and relieve myself.

My friends are sitting a few yards away resting. I look for something to wipe myself and find a smooth stone a yard and half away. Toddle towards the stone, crouch and wipe. Identify another stone and wipe again. And so with a third. Walk towards the eucalyptus tree and cut a small branch and perform the last wipe. What a relief. Straighten up, pull up my underwear and pants and stretch my body. Pick up my gun and military belt and put them on. My friends are ready to move.

I say, "Wow. That was good."

My friends grumble their assent and we rush towards the next navigation point. Marching swiftly. The sun travels west.

The dozens of miles we covered each day of the navigation drill during that week were preparations for the more demanding night-time challenges.

Galilee—By Night

...and Joshua went that night into the midst of the vale.

(Joshua, Ch. 8, v. 13)

During night navigation drills, the navigator needed to memorize the trail by heart. Everything is different; the sights, sounds, smells, sensations, temperature. The night navigation trails are longer than the day trails and, in most instances, we could not complete the march during the night. Dawn would find us still on the trail and illuminate our aching bodies.

The transition from resting comfortably in the truck to preparing for the night march was always a challenge. The inner battle commenced while we were still in the truck, anticipating the moment we would get off and begin marching. I felt my body rebelling and can still remember the sensations of *I don't want to* and *I want to doze some more*. All are descending from the truck and I join them. We make sure all of our equipment is in place. Checking each item to make sure it is tightly packed and won't rattle. We drink water, find north and the direction we are heading. Among our team of three, we recite a description of the first mile to each other. I can revisualize the entire navigation trail from beginning to the end point. Can see myself arriving at the end point before I take my first step. The end point is identified by the surrounding hills, asphalt road and the most important item—the D400 truck. We descend from the vehicle only to climb back on again at the end of the night. Actually... the next morning. We are all focused on reaching the D-400 which holds our dry clothing and rubber mattresses. We envision resting on our backs, closing our eyes, protected from the rain drops within the relative darkness. The smell is not lovely but familiar.

Our team was comprised of Yadi, Dany and myself. During the first two days of the night navigation drill, we realized that around 2 a.m. we lose energy and concentration. Yadi and I talked about it and decided that we needed to move as fast as we could. Dany agreed with us. We marched swiftly, covering as many miles as possible before 2 a.m. At that hour, our mental acuity became foggy and we wanted to be as close to the trail's end as possible. Yadi and I felt a strong desire to finish the trail as quickly as possible. Dany, who was in great physical shape, carried us with his energetic spirit on parts of the trail where we got weaker.

Yadi and I look at the compass to ensure that we were starting out in the right direction that would bring us quickly to the first navigation landmark.

I question, "We march towards the flat saddle ahead of us. After 450 yards we cross a wadi and then we climb according to the azimuth to the saddle"?

Yadi approves by nodding his head and Dany joins in. Then, like racing horses we rush ahead. All our physical energy is concentrated on the forced march. My mind propels me forward with repeated directives of *Forward. Forward. Forward.* I am fully focused on the next navigation point. We walk as rapidly as we can to close the gap to that point. I spur myself on and am spurred on by my teammates. We race against the other teams. *Who will arrive first to the truck?* Nobody openly acknowledges the race but it exists, and drives our navigation accuracy and speed.

We accomplish the first 12 miles, overcome them eagerly. *We are doing well* the voice in my head says, impelling me forward. We drink some water and take a short rest. Now we begin the more difficult third quarter. I describe the next section of the trail to Yadi and Dany and get their approval. We remind each other that we are about to enter the "exhaustion" stage of the drill. A time when we become apathetic and hazy, lured by the night's siren call to sleep. We remind ourselves to "stay hungry" and "devour" the next portion of the trail and look at each other eyes to affirm our shared commitment. "OK. Continue with energy." We stand up and rush onward. The fast pace energizes us. Dany and Yadi's rapid pace spurs me on and my pace in turn, energizes them.

All three of us are invigorated and challenge ourselves to continue with the spirited pace. My shirt is wet and sticky. The song "We are walking, Hi, we walk, Hi," enters my mind and lightens my body. It stays with me and rejuvenates me so that my legs don't feel tired. I feel joyous and alert and sleep is far away. My nose is running. I lift my right thumb to the right nostril and squeeze to clear my left nostril. Place left thumb on the left nostril and repeat. Wipe my nose on my right sleeve. *I will breathe easily through my nose for a while.*

I focus on the trail's next navigation point, glad for the quick pace. It is 3 a.m. The darkest time of the night. We are engulfed by the murky depths of opaquest blackness. At this point, little oxygen is reaching our brains and we fight to stay alert. However, this night our sensations are different from previous nights. Although I did not drink coffee or any other stimulants, the shared encouragement of Dany and Yadi stimulates my body and mind. Our brisk pace enables us, for the first time, to complete the navigation trail before sun rise and in better than usual mental and physical shape.

We begin to understand the connection between "being fit" and the ability to complete the navigation trail quickly and with precision by supporting each other. We are building the navigation communications of the crew, communicating with ease and often without words. Our shared language is anchored in the steepness of hills, the terrain temperature, distance traveled, rocks, muscle aches, north, south, azimuth degrees, north pole star, fog, mist, moon, well and water hole. And the nonstop questions "Where am I now?" "What is the distance to the next navigation point and what direction will take me there? The fastest?"

The Cotton Field

> *For all his days are pains, and his occupation vexation;*
> *yea, even in the night his heart taketh not rest.*
>
> (Ecclesiastes Ch. 2, v. 23)

Fourth night. We are marching energetically knowing we are close to the final section of the trail. We know that within one hour dawn will break. We are tired. Our legs ache because of the dozens of miles covered in each of the three previous nights. These marches didn't end until well after sunrise. Towards the end of the fourth night we find ourselves in a huge cotton field, extending from Mount Tavor to the asphalt road that connects Afula to Kaduri School where Yzhak Rabin studied. The field seems empty since the plants were cut leaving one to eight inches high, sharp stalks. The big toe on my right foot is hurting and I try to ignore it. I urge myself to march faster towards the trail end. Walk as fast as I can.

My boot hits a cotton stalk every few minutes and I try to ignore the pain. The number of stalks striking my boot increases. One stalk pushes against my toe. Piercing pain! I jump as if bitten by a snake. The level of pain is unexpected and a loud curse escapes my mouth.

Yadi turns to me with a questioning look, "What happened?"

I raise my hand signifying "Nothing." We continue. Another stalk traumatizes my toe and the wave of pain penetrates like a hammer pounding a nail into me. I managed to stifle the scream, and just a tiny whimper escapes me. I am afraid to continue walking normally. The anticipation of the next stalk piercing my toe fills me with dread.

I decide to try walking with my right foot angled so that stalks will pierce the fleshy part of my sole, or my little toe, but not the injured big toe. The awkward walking motion relieves the tension of anticipating a new injury to my toe. But now I have other things to take care of. I concentrate on my twisted right foot and feel the painful toe acutely. The stress of tensing my right ankle bothers me. As I straighten the twisted ankle, a stalk hits the toe again and the shocking pain forces me to twist my foot again. I straighten my ankle, twist again and another thorn pierces my boot. It is so painful that tears come to my eyes. Continue to walk at a short distance behind my friends and center my thoughts inside myself. Concentrate on holding the twisted foot to the left. Jump with each stalk hitting my boot.

The night stretches out before me like an eternity. I move deeper inside myself and away from my external existence. Feel the emotional ties that connect me to my two fellow crew members but stay deeply immersed inside myself. Sometimes a thought about the sharp pain that I'm anticipating and then … shocking pain. Attempt to ignore it by digging deeper inside myself. I am encased in a secure cocoon and the external environment has no effect on me. I have no desire to open myself to the sensations of this prolonged nightmarish march. When I venture outside my cocoon, I am all hurting right foot. Sharp piercing pain in my big toe. Go back inside. With no warning, a hint of a light breaks through the dense night. A wave of joy fills me. I let down my guard and am rewarded with another direct hit from an ambushing sharp stalk. On we go. I can see the curve in the asphalt road signifying the trail end and the eucalyptus trees behind it. *A bit farther. A bit farther.* The night sky is clearing, slowly fading into a soft pink color. We march faster. I take another hit to my toe and whimper. Finally arrive at the end point. We sit at the side of the road, and I take off my boot. What a relief. Exhale a long breath and remove my sock. The big toe is inflamed and enlarged. The truck arrives and we climb in. Another night of navigation drills has ended. I have conquered this nightmarish trail.

Last Galilee Navigation Night

On Thursday, before the last Galilee navigation night, we found ourselves a few miles east of Mount Tavor, south of Kaduri School. Each drill started at the end point of the previous night's navigation night trail. There we had a truck, water, IDF food, basic cooking equipment, 1:50,000 navigation maps, and some personal gear. At the end of the night trail, we all collapsed exhausted and fell immediately asleep. Few hours pass and we started, with some reluctance, to wake up. We tried to find shelter under the large eucalyptus tree. Despite its abundance of leaves, it did not provide much shade. The sun climbed higher in the sky and the heat rose accordingly. In contrast to the lack of shade, there was a profusion of flies buzzing around us. The combination of exhaustion, minimal shade and swarm of flies, was enough to defeat any soldier, including a Unit combat soldier.

It seemed like the flies deliberately decided to wake me up. I slept on a rubber IDF mattress that had a large dry sweat stain in the center and small rips in the cover that revealed its spongy insides. Adding to the décor were a few dark stains. Grease? Oil? Food remains? They were spread all over. As protection against the flies, I covered most of my skin with my IDF shirt and pants while leaving my legs free to get fresh air to heal from the demanding night walk. My hat covered my face and protected my eyes from the blinding sun light.

My feet and palms were fresh meat for the flies to lick and bite. Most of the time I was unaware of them landing on me until the moment one started to bite. Without thinking, I would scratch one foot with the other to get rid of the fly. I did not want to put on my socks and block the fresh air, but after a few more bites I placed another shirt over my feet. That area is protected now. My face and neck are still exposed, however. My ears also drew the flies like honey or jam, although I had neither there to lure them. Do my ears, unwashed for four days while walking dozens of miles each night, smell attractive to flies looking for a mate? I am not concerned by the biology. Just let me rest. My mouth and nose are open. My humid warm breath attracts the flies. Half asleep, I would lightly caress these areas to shoo away the flies.

The heat continues to rise. I am so tired and dying to sleep. More flies are landing and biting. I find a T-shirt and wrap it around my head, leaving a small slit for breathing. Peace. I fall asleep again sinking into a sweet soft slumber. I get a few minutes of rest until a bite from a stubborn fly, having a late breakfast, wakes me up.

I sit up and try to open my eyes and breathe in the morning air that has heated up in the last hour. Still fuzzy, I put on my sandals and move towards the water tank in the back of the truck as my eyes slowly open. Drink some water and wash my face. Drink some more and stumble back to the crew area. Take a bite from a cucumber. Enjoy the wet-green freshness. Continue to wake up ploddingly. Pour soapy water into a container and soak my inflamed big toe, knowing that within half an hour the swelling will subside. In this pastoral atmosphere we have two easy tasks: eat and rest as much as we can. Ah, yes, and another more exciting task—study tonight's navigation trail. Dany and Yadi studied the trail together while I was napping. They were alert for a while.

As preparation, we would quiz each other about the details of the trail. At first, one of us would ask and we would describe the trail while looking at the map. As we got better acquainted with the trail, we could describe it while occasionally glancing at the map. Finally, we would be able to describe the trail by heart, without checking the map. I told Yadi that I would quiz him about the trail in the afternoon. Yadi passed me the map and I questioned him for 20 minutes, commenting on the trail sections requiring attention. Dany listened to us. As we completed the exam, I asked Yadi to question me. Yadi was surprised and asked, "How can you know the trail by heart?" I responded that I thought I could remember it. We were both amazed that I could visualize the trail with nearly no mistakes. I managed to picture the trail in my mind. I questioned Dany again and realized that I knew the trail well.

I found a shadier spot, covered myself with a thin shirt as protection from the flies and dozed. At some points I actually fell asleep but the flies woke me every few moments. Hours passed and the sun slowly moved west. The heat blanketed the field. In a few moments, darkness will conquer light and we will hurry to begin the last navigation trail of the week.

That night, we practically ran the trail with utmost precision. Traversed the first miles north towards Golani Junction, leaving Kibbutz Beit Keshet on our left. Crossed the low ranges and passed the paved road extending from Afula to Tiberius. Leaving the junction south of us and continued heading northeast. The difficult part was going west of Mount Nemra and descending into the wadi that would lead us north of Doron Hill. We knew that section would be the most challenging navigation since we had to walk on a range with no significant landmarks to anchor us and verify that we are on the correct trail. The direction was clear and there were no indications that we might be off trail.

As we descended into the wadi we were able to verify that we were on track by finding the two water holes that were navigation points for us. The wadi continues downward, heading into the Sea of Galilee and we run east. The distance to the end of the trail is shrinking since the last night trail is relatively short. We concentrate on upping the already rapid pace and extend our stride. Our team is the first to reach the end point at Villa Melchet and the waiting truck, the restful "red-lights-cave" at the end of our week of Galilee night trails.

The Crew members stand together at the beginning of the training track. Everyone is pushed forward by his competitiveness, the drive to prove, the fear of shame and dismissal. Extreme ambition. We all are very competitive and each of us attempts to outdo the others. We all know the rivalry exists but no one talks about it. Instead, we talk about the training missions, news and stories from home. Each one of us wants to do everything in the most accurate and impressive way, to excel and earn the highest score in the crew. As in life, with each task there are some who do better than others. Some of our competitiveness expresses itself in assisting others while performing the required tasks.

Each person who assists a comrade, "earns points" that are not acknowledged out loud but are noted. Points are also earned by the one who accepts the assistance. And each crew member participates upping the score without discussing it. We rarely talked. Until today, most of us shared very little. We did not clarify what we thought or felt. We were motivated to keep going, keep walking and survive one week at a time. Our aim was not to be dismissed. In the next task, the next day, the next month. We ignored our exhaustion and pains. We supported and assisted each other. And continued to perform.

Desert Challenge

And the angel of the LORD found her by a fountain of water in the wilderness, by the fountain in the way to Shur.

(Genesis, Ch. 16, v. 7)

Water is scarce and valuable in the desert. To this day. We could navigate in the water-starved Negev Desert (the southern part of Israel) only after we had practiced navigation in the north. Walking in the Negev is challenging. It is a dry, arid, dissected region that supports little vegetation, with narrow passages through cliffs. There are few villages and roads. Navigation in the Negev desert is planned for the winter, the coldest time of the year. During summer, the heat is so intense that it is nearly impossible to cover dozens of miles, day after day.

We traveled south by truck on a wintry day and began our walk from the center of the Negev, heading towards Eilat, the southern port of Israel. Each group walking the trail consisted of three soldiers. The Negev scenery comprises vast open stretches that allow us to periodically see the other groups walking the trail. At evening, when all the groups had reached the trail's end point, we prepared a meal. During the winter of 1972, we walked five long challenging days in a row, and our brains and bodies got adjusted to the strain.

To protect our feet, we had several tricks. The first was to cover our soles from toe to heel with an adhesive bandage, practically adding another dermal layer so that our boots rubbed against the bandage rather than our skin. This trick did not allow the skin to "breathe," but it delayed our feet hurting with each step. Nevertheless, blisters do eventually appear. On the third day, my feet were covered with blisters.

For additional protection, we would put on a very tight, thin sock and a thick sock on top of that, careful to have the seam line in an area where the boot fit wasn't too tight, to avoid creating another sore spot. Usually, we placed the seam over the toes, but never at the tips. Doron would smell his socks before pulling them over his feet. He would mutter: "If I'm not yet running away from the smell, it means that I can walk with these socks for another day." We laughed.

We diligently trimmed our toenails so that there was no extra pressure at the nail bed. A long nail becomes painful especially as we went downhill and each step halted the body's forward motion, causing the nail to collide with the front of the boot. Long nails became a nightmare that taught soldiers to cut their nails at least once a week.

We were "crazy" about crepe soles. We developed "crepe cult" practices that enabled us to use these flexible soles for most of our walks. Traditionally, only paratroopers were allowed the privilege of using crepe soles during their IDF service. Crepe sole were a distinguishing mark of an elite soldier in any Israeli crowd. As soon as we identified any tiny wear to our boots or crepe soles, we would rush to the equipment hanger for hasty repair. We managed to keep a stash of crepe soles so that we would always have back-ups. 'Crepe' soles are not suitable for muddy terrain, but their ability to absorb shock is superb. Before each week of navigation drills, we made sure that our crepe boots were intact and ready to go. We carefully checked that the soles were firmly attached to the boot leather to ensure that not even a smidgen of a crack existed. If we identified a small crack, we would glue it closed right away with contact glue. If the crack was larger, we rushed the shoe to the maintenance equipment hanger for repairs.

By the fourth day of walking, we had covered over 80 miles. I was part of a navigation crew with Amnon and Dany. They both had amazing physical stamina. We began that day before 6 a.m. after drinking a lukewarm liquid that passed for tea. Around 7:30, we took a five-minute break to drink from one of the canteens. We looked east towards the Arava and then south, seeking to identify landmarks along the long navigation trail that lay ahead of us. All three of us were exhausted. My feet were in worse shape than my friends. At the end of each break, I was the first to get up and start limping down the trail. My friends waited until I was a few hundred yards ahead and then followed, catching up with me easily and then slowing so we could all walk at the same pace.

To deflect the pain, I grit my teeth and continue to limp. I know that after about 500 yards the blisters will warm up and the pain in my feet will lessen...and then move into my leg muscles. I limp the first few steps, cursing intensely with each step—"Kosamak!" I keep my voice low since the pain is all mine. I shift quickly from foot to foot to avoid putting my weight more than a second on one or the other. My feet shoot messages of excruciating pain at the speed of light to my brain. Amnon and Dany see that I have gotten far enough ahead, get up and walk to join me, laughing at my funny duck waddle. I join their laughter. A strange yet effective way to suppress the pain. On we go. The pain in my feet moves to my muscles. As I walk, this pain abates.

"The walker of the land conquers it."*

Conquer?

I do not conquer. I endure.

* A common saying in Israel.

I pay walking taxes. Sweat taxes. Blisters taxes. Muscle ache taxes. Dry mouth taxes. Blurred vision taxes. Extraordinary effort taxes. Struggling to catch my breath taxes. Vast range of pains taxes.

I do not conquer and am not conquered. Just a walker who was given permission to pass through and take in the sights.

At 10 a.m. we took another break. At that point we had covered about a third of the trail. I felt ready to "call it a day," but we were required to reach the end of the trail, together. Hour after hour we continued to march in the arid Negev Desert. Descending from hill to wide wadi, climbing again, following the trail. The sun was reaching its zenith and the day became hotter. The bright light hurt our slitted eyes and a few wind gusts coming from the west chilled us while forcing small grains of sand into our ears. In the afternoon I turned my hat around to protect the right side of my face from the sun. Three gray desert falcon were hovering above our heads. I touched one of my friend's shoulders and pointed up towards the birds of prey. We both burst out laughing at the shared thought: "We might look like roadkill, but we are not dead yet."

At 5 p.m. we were still a few miles from the end point on the asphalt road. We looked at each other. Walking. With no words, just communicating with a look, we agree to give up the next rest break and increase our marching pace. We will rest at the end of this trail, two miles ahead, not now. I focus on our faster gait, ignoring my pains. The light from the setting sun, blueish red, is absorbed into a dark desert night. Twenty minutes later we see the red tail lights of the truck at the end of the trail. Fifty yards before the end we can no longer wait and take off running to reach the finish. A unit tradition. Then we take off our boots and collapse.

I start the last day of the south navigation trail with my body still aching. I get to know pains in muscles I never knew existed. My feet are burning. Start walking. The wild, untamed desert scenery surrounds me. Words from the poem "Bemaale Akrabim"—"In the Scorpion's Ascent"—by Aein Hillel, run through my brain.

Abruptly
God exploded into my eyes...
Broke before cliff mountains
Descending to the Arava Valley.
... Exploded Desert God...
I did not think, just inhaled molten lead...
And you had handed me the cold steel to kill your frail creatures...
And behold I am going to kill worthless ones like me...
And my soul shatters to become eternity...

Beauty fills me. Tears come to my eyes. The walk continues.

We look at the map identifying the trail climbing from the east towards the Solomon Mountain top where our next navigation landmark is located. We chose our trail and start climbing. For some reason the contour lines marking altitude do not make sense. We look at the map again and all of a sudden understand that the gap between the lines indicates a height of 20 yards rather than the 10 yards typical of the other maps we've been using to navigate. I realize that the ascent will be much more demanding. As we finish the first uphill stretch, we feel the difference painfully in our bodies.

We continue our climb towards the top. The ascent is becoming increasingly steeper and the trail narrows until we are walking on a razor-thin edge. On either side there is an abyss and we are suspended in space. The rock face changes from brown chalk to magmatic black rock. We continue our ascent. At times there is a narrow saddle that forces us to descend a few yards and then climb again. We are less than a mile from the top. Any glance to either side makes me dizzy. The knife's blade edge that is the trail has become even more perilous. My foot slides on loose rock and I fall on my butt to steady myself. The swinging rifle jerks me off balance for a moment. Only a few hundred yards to the top. My fear of heights is getting worse. I move forward on my hands and knees or slide forward on my butt. Amnon and Dany are calm, moving back and forth along the knife blade with no effort or hesitation. Two mountain goats enjoying a sunny day. And me? Scared to the depth of my soul. Each step threatens to drop me into the abyss to either side.

At one point, my right foot slides and a rock tumbles down the slope. As it gains speed, it sets off a small avalanche of stones. I sit and breathe. An angry shout comes from far below.

"Its Yehuda's team," Amnon mutters, nonchalantly.

We reach a passage that forces me to cling to the rock and detour to the right. I don't see how I can get around it because my machine gun will throw me off-balance. My fear overcomes me for a moment. Each look down amplifies my acrophobia. I know that I have to move forward to complete this trail. I swallow my pride and ask Dany to take my Uzi. He steps beside me and takes it. In a second he is on the other side of the intimidating rock. I ask for more help. Amnon comes towards me. He is not a big talker but always leads in making the correct command decision. He developed that capability during his experience at sea while growing up in Kibbutz Magan Michael. Always forthright and businesslike. Always courageous. Always with great caring and concern camouflaged with cynical dry humor.

Amnon directs me with his calm, assuring voice and I slowly move forward towards his outstretched hand. I rely on his voice and try to ignore the abyss on either side. I all but hug the rock and concentrate on the reassuring voice of Amnon. Another step and...

Phew, we are on the other side of the rock. A few more yards and we are on the top of Mount Solomon.

We sit at the top and look toward the end point of the trail. We decide to go down towards the paved road on the southern side of the mountain, which is the quickest way down. We are well aware of the sun setting in the west. We do not want to be on the mountain when darkness falls. Moving down, we enter a narrow wadi and come to a small, dry water fall. Amnon and Dany detour around it in seconds, but my fear of heights overcomes me again. I ask for help. Amnon patiently guides me down the cliff with his calming voice. Going down the wadi for dozens of yards, I slow my breathing and feel calmer. I almost ask for my Uzi back, but we reach another waterfall ledge. Again, Amnon and Dany hop down while I am paralyzed by my fear of heights. Make the descent with Amnon's help. We continue climbing down. Amnon and Dany move quickly. I plod along, getting stuck at every passage that presents itself as an "inviting" abyss.

When we reach a slope of small, magmatic rocks, colored black and brown, we can see the paved road below us. All we have to do is crouch low and slide down. I hope that we will not cause a rock avalanche as the dislodged rocks pick up speed downhill. We slide on our behinds at a steady pace as if heading down a chute. We control our descent by sticking out our legs to reduce our speed. As we near the paved road my fear of heights diminishes. I can see to where I would land if I fell and the distance is not that great. We reach the last cliff before the road. Amnon and Dany go down easily. I slowly catch up. Almost surprised when I am on the paved road.

A huge sense of relief.

I am very glad to leave Mount Solomon behind and the paralyzing fear of heights that accompanied me on this trail. It was enough simply to visualize myself on the knife edge to paralyze me. Now, the dizzying abyss and the frightful sensations are over. My nervous system is still on high alert. I'm ready to drop flat to the ground to avoid rolling downhill.

We are on the paved road. Breathe and accept it. Breathe and absorb the reality of standing on a simple paved road. "Yes, on the road." I'm filled with a tremendous sense of relief, joy and gratification. The intoxication of being freed from my fear of heights combines with the great satisfaction of having successfully completed the week of navigating Israel's southern trails.

Tricky Hebron Hills

And the people stood afar off; but Moses drew near into the thick darkness.
(Exodus Ch. 20, v. 17)

Now, that we had accumulated much navigation experience, we were required to face a more challenging task—night navigation in the Hebron Mountains. The geographical lines of the mountains in the Galilee are clear and simple by comparison. The mountain range at the center of Israel commences at the Galilee and continues southward towards the city of Beer Sheva via Nablus, Jerusalem and Hebron. The general geographical structure is clear—on the west the Mediterranean, to the east the Syrian African Fault and the Israeli mountain range located in the middle. The brooks flow in straight wadis from the watershed line towards the Mediterranean Sea on the west and the Syrian-African fault to the east.

In the Hebron area, the formations of the wadis are more complicated. Many wind around and split, requiring a thorough study of the trail. Since we knew that, we began the task of mapping and memorizing the first two-night trails a week ahead of time.

It was important for me to learn as many details as I could. Just remembering the basics of the trail would be sufficient, but I wanted more. My desire to excel bordered on "showing-off," which can create problems. When the trails are short, the ability to show-off with knowledge of a special tree nearby is not harmful, but with trails exceeding 25 miles, excess details interfere with the basic navigation requirements of direction and distance, running the risk that they will become blurred and fuzzy.

When Doron, Ben Ami, and I started walking, I felt that my head was stuffed with too many details. The beginning of the trail was enjoyable since we could recognize all the landmarks we had memorized during our preparation and "say hello" to each as we passed. After about 20 miles, we climbed several 100 yards and transversed the elevated portion of the mountains. Exhaustion affected our minds and bodies and fatigue began to influence our judgment. In addition to muddling the trail details in my mind, a thick fog enveloped the Hebron Mountain and the temperature became colder and colder. As long as we walked, it was bearable and we did not suffer from hypothermia.

We entered a small wadi and tracked right for 200 yards. Climbed 600 yards uphill towards the mountain ridge. We expected to find a water hole there. Although we located it, the water hole didn't appear to be where my study of the trail indicated it should be, 50 yards south of the ridge. Instead, it was situated north of the ridge rather than the south part. The fog thickened. The moon sank behind the horizon and disappeared. We were surrounded by foggy darkness.

We stop. Something is not right. I search my memory for clues where we may have deviated from the trail. Figuring that I probably turned right instead of left, and then walked 200 yards, I conclude that we are likely 400 to 500 yards from the trail and walking parallel to it. I try to recall what landmarks are to the right of the trail but can't reconstruct the details. There are too many small, winding wadis. I remember that the most prominent nearby landmark is the watershed we will cross in about three miles. From there, we can see the end point of the trail in a wide wadi.

I tell the others, "We'll climb for another three miles to arrive at a location where we can see where the trail continues. There we should be able to identify where we are." I laugh to myself thinking, *In this fog it's unlikely we will be able to see anything.*

A look at my watch tells me that first light is less than one hour away. We march uphill. I am breathing heavily and sweating profusely. We have covered dozens of miles, but many miles still lie ahead of us. A few minutes before dawn, the small voice in my head tells me that night is about to be transformed by morning sunlight, but my mind is still immersed in the night's onerous reality. We continue to climb. A short distance before the watershed line, the skies become clearer and first light breaks. The night fog is melting, conquered by the strengthening light. The ridge line ahead is clearer and we are very close to it. We cross it and find a large rock to sit on. The deep blue of the sky is colored with cautious azure, pink and red just before sunrise.

We see a wide wadi ahead of us and know that if it marks the trail end point, a truck will be waiting for us at the intersection with the paved road. I look through my binoculars in the direction where I think that might be and make out two weak red lights. Excitedly, I hand the binoculars to Doron, who has "hawk eyes." He looks and confirms that it is the truck. We erupt with joy, quickly gulp down some water and gallop the remaining four-and-a-half miles down the wadi to the truck. As we arrive, we get rid of our military belts. We all take a short nap to renew our energy, which we will need later to study the trail for the following night.

After covering more than 50 miles in two nights, we are nearing the final segment of the third navigation. Aching muscles and joint pains have become integral parts of our existence. Doron, Ben Ami and I walk down a moderately sloped path through a wadi heading west. To our right there is a steep slope and on our left the wadi's river

* Powdered soil, common in the southern part of Israel, that hardens with exposure to water.

bed is filled with rocks. Cracked loess soil* lies between the river bed, the steep slope, and the trail that stretches along the wadi's bank. Like the Bedouin goats, we walk on this comfortable loess strip.

The wadi changes direction and we stop. The direction seems right, but it appears that the wadi is starting to rise rather than descend. We have been walking down this wadi for more than half an hour and all of a sudden it goes up? Can it do that? Impossible. Yet it seems to. We all look at the compass again and check that the direction is right. I ask Doron and Ben Ami to wait for me. Walk forward about 50 yards and look back at my friends. Now I'm no longer certain that the wadi is going up. We decide to continue. After another 200 yards, we are sure that the wadi is going down. *What a relief. What were we seeing? An optical illusion?* Perhaps.

On we go. The wadi is getting wider and the direction is clear. We increase our pace. The walking is monotonous. There are many more miles to go before we reach the trail's end. I lead, Doron follows me and Ben Ami follows. I glance back every few minutes to make sure they're following me. At some point, I stop and propose that we move faster so that we will reach the finish during the early morning hours. We quicken our pace and keep two to three yards between us. The trail is clear and we march fast. Every once in a while, there is a gap in the trail where the water has eroded the bank and the trail has been redirected to avoid the gap.

I recall erosion in the "Small Crater" of the Negev where water undermined the soft sand layers below the hard chalk rocks until they fell and were shifted by the occasional floods. Here, where the wadi curves, it ate into the Loess bank and created holes. When the wadi flows straight again, the trail straightens, too. I look back and see that Ben Ami is falling behind, but I keep going. The fog is rolling in. Fatigue clouds my previously solid knowledge of the trail's details. As we tire, performing simple tasks becomes more difficult. The fog around us seems to penetrate my aching brain, but I am not too concerned since the navigation tasks are not demanding. Nevertheless, the direction is clear and simple. Each of us sinks into our own separate bubble moving with the rhythm of our steps. The trail swings along the wadi's bank. I look back. All is well. I keep marching.

All of a sudden, I hear a choked shout. I look back.

Where is Ben Ami?

I stop. My body tenses. I'm fully alert now and concentrating on my surroundings I ask Doron, "Where is Ben Ami?"

Doron wakes up. "I don't know."

Both of us walk back.

"Ben Ami?!" I shout loudly. I think of Ben Ami who likes to be the counterpoint in our crew and take opposing positions that sneer at our arrogant pose as "Matkalists."

"Here," I hear a few feet from me.

I look down and see Ben Ami inside one of the wadi bank cracks. I reach my hand down; he grabs it and climbs out of the hole. He is OK.

On we go. I imagine myself falling into this hole and the tension release causes me to laugh out loud.

"What's up?" Doron asks.

"Nothing," I respond.

We continue to walk in silence. At times my boot strikes a hyssop shrub and the pleasant smell fills my nostrils. We march forcefully through the thick darkness just before dawn.

First light. I take a deep breath. The transition from the pitch black to deep blue that signals the approaching dawn fills me with joy. My body and heart are elated by the increasing light of morning and a new day.

We cross a field of cucumbers covered with thick dew. The cucumbers look bright. Hunger suddenly gnaws at me. I bend down and pick up a cucumber. It feels coarse as I bite into it. Strange, what kind of cucumber is this? I take another bite, chew, and look at it. The taste is different from the look. Aahh...it is a zucchini!

I take a few more bites of the fleshy part of the zucchini and throw away the rind. Enjoy the taste of the wet pulp soothing my hunger pangs. We stop at the field edge and drink water. In a few miles we will reach the truck that waits for us at the end point.

Off we go to finish our last night navigating in the Hebron Mountains. Tonight we will descend the southern slopes towards the wide, dry Beer Sheva riverbed and march along in it to the end of the trail.

We arrive at the Lahav Forest, close to my Kibbutz, Mishmar Hanegev. We can see the Tel Grar archaeological site. So many times I was there with Lotan, Meir, Yishai and others on foot, by tractor or jeep. Remember trips with Meir to pick figs and sabres (prickly pears). Meir and I started with a stick and an empty can and developed a system to cut the edge of the sabres leaves with one blow from a sharp knife to separate the fruit. We filled a bucket, used the high-pressure hydrant near the tractor garage to wash off all the thorns, and brought the fruit to our class mates. We sat on the grass in front of Lotan and Yossi's room, some of us swinging in the hammocks, and had a summer feast. Several times we sneaked into the kibbutz kitchen to snitch ice cream or popsicles for desert.

The last trail is shorter and simpler for navigation ...relatively. After a brisk march that covers dozens of miles, we reach a navigation point on a hilltop. We are all tired and thirsty. We decide to take a "2 by 2" break, meaning each crew member stands guard for two minutes while the other two sleep. The soldiers at the beginning and end positions have it best because they get four minutes of uninterrupted sleep while those in the middle have to settle for two brief sleeping periods. We go down into a small wadi bed for protection from the wind. We nap on the warm ground, using our military belts as a pillow. I am in the middle. Ben Ami finishes his turn, wakes me and falls asleep. I look at my watch, yawn, and move a few steps away to piss. Two minutes pass. I wake up Doron. He sits up and I go back to sleep. After what seems like a very long time, Doron jumps up and wakes both of us.

"A donkey licked my face" he exclaims in a loud voice.

We look around and don't see a donkey or any other animal. I look at my watch and note that over five minutes has passed.

I get up and we start to walk. I lead on a steep slope. Our vision is limited by the fog and the lack of moonlight. I am close to the edge of a cliff. I stop for a moment and realize that my visual field is distorted and can't be trusted. I move a few feet to the right along the cliff's edge and see no path for descent. I shift several dozen feet to the left. The cliff's edge does not change and I can't find a path down. I try to remember what the map showed. My vision is blurred and the cliff seems frightening. How many yards to the base of the cliff? Five? Twenty?

I am apprehensive. Doron and Ben Ami are behind me. I stand again on the cliff's edge and look intently into the fog, attempting to gauge the distance down. Fear knots my stomach and constricts my chest. I decide to shimmy down. How? I don't know yet. I ask Doron to stand alongside of me so that, if needed, he can help me climb back up. I sit on my butt and let my legs dangle over the edge then turn my body so I can descend into the abyss. My fear rises and I am halfway down before my boot touches solid rock. I put my other leg down and it touches rock as well. Both boots are firmly planted on a solid rock bed. The abyss is barely a yard and a half deep. What a relief. I take a deep breath. I signal for Doron to join me and Ben Ami follows. My crippling fear of heights is evaporating. We walk down the hill heading south.

Doron continues to tell us about the donkey that licked him and we all laugh. We create variations of the story that will become another piece of crew legend.

Doron says, "Although I was sitting up, I fell asleep anyway and all of a sudden I wake up feeling a donkey licking me." All three of us laugh, imaging what it would feel like to be licked by a Bedouin's donkey.

"Doron, please bring the donkey along so we can ride it," Ben Ami says.

We laugh again.

And so our three-man navigation crew walks into the wide Beer Sheva dry riverbed still laughing. To the west we see the huge halo cast by Beer Sheva, the largest city in the Negev Desert. The light overshadows us on our last miles of our last trail during Hebron navigation week.

Azimuth Navigation

Turn our captivity, O LORD, as the streams in the dry land.
(Psalms Ch. 126, V. 4)

Azimuth navigation*. Seems simple. All one needs to remember is direction and distance. Only a few stretches, 10 to 12, to remember. Also, one needs to learn where the landmarks are located along the trail. After the navigations in the Galilee and the west bank with their high mountains and steep slopes, the flat region of the northern Negev seems easy by comparison. For me it is close to home, a few miles from Kibbutz Mishmar Hanegev, where I grew up. The fields and the smells are familiar. The navigation trails pass through points I visited on trips during my teen years. I visited Tel Grar and Nachal Grar Wadi many times.

To make the task challenging, Azimuth navigations will allow few breaks. We will walk day, night, day, night and another day in a row. We began navigation during the day. Sunrise finds us starting out on the long trail. At the end of the day trail, a truck will be waiting for us with some warm food. We will eat, study the night trail for an hour or two and then begin to walk again.

Our navigation squad consists of Lony, Yermi and myself. With Lony, I feel very comfortable. Time spent with him is like entering a Yeshiva room and discussing an interesting Gemora chapter.**

---

* Navigating by using a compass. Azimuth is expressed by a degree between 0 and 360. For example, "going on Azimuth 90", means going east.

** A rabbinical commentary on sections from the Bible and the Talmud.

We both learn the trail quickly and, as we walk, compare our knowledge of the trail details. Lony and I had our routine, and Yermi counts on us to lead the way as he follows. Our other crew members know that during the middle of the night, Lony and I will start discussing the next portion of the trail.

It goes like this: "Now we will go in the direction of 183 degrees for 700 yards and pass a ruin on our right."

"After another 700 yards there is a road and the ruin is located 15 to 20 yards to the right of that road. I think that if we go on 184 degrees, we will get closer to the ruin by 10 yards."

"I bet the distance is a few yards longer than 700. Let's measure with our steps."

"OK, six minutes on the watch."

"5:45, but don't drag your feet."

"I'll take the lead and even at our brisk pace, it will take six minutes. Remember the small wadi after 370 yards."

"I remember. It's small. Let's go!"

During the 30 seconds it takes us to discuss what lies ahead, we share our knowledge and opinions, quantify the facts, and bet on the results. When we arrive at the landmark, we review the results and best guess wins points. During the night, everyone's accumulated points are counted.

Navigation, above all, is a mental exercise. I constantly double-check with myself if I am "on the trail" and test if I might have deviated. I search for signs to verify that I am staying on course. Most important is to maintain direction and distance. Direction is the highest priority. I measure the distance with "double steps"—two steps at a time.

At times, there are clear road signs and I don't need to do that. The navigator brain works like a scientist experimenting. I assume that I am at a certain point, progress to the next point, and search for clues to prove or disprove that I am on track. All this happens while still moving. If too much time passes and I can't find evidence supporting that I am on the right path, I start to consider points where I might have gone off the trail. If I deviated to the right, I will hit a wadi flowing from northeast towards southwest. If I deviated to the left, I will reach the bottom of the hill in 500 yards. I wrack my brain to look for signs that will prove I am on the right track. I try to remember what I saw at the map.

Usually, during the night hours, I could not extract the additional details I memorized before we set out on the trail. I would make mental notes of small details like the halos from the lights of villages and towns, a prominent hill with a saddle and its orientation, and the flow direction of a large wadi relative to the navigation trail. Every detail that can assist me in pinpointing my location on the trail would be crammed into my head. As we improved our navigation skills, we learned to ask each other certain key questions before going out on the trail. Answering these questions enabled us to identify mistakes and correct ourselves quickly when we deviated from the trail. Despite the in-depth study, I constantly ask the question "Where am I?" on the navigation trail. This leads to numerous "smaller" questions, such as: Is this hill 800- or 900-yards' distance from me? Will the saddle appear clearly or is it too flat? How many yards before the small wadi rising east-south-east appears? These types of questions keep me focused on the trail and constantly verifying that I am on the right track, accurately estimating arrival times to the next navigation point and correcting mistakes on the move. At some navigation points we stop to take a short drinking break. It is easier to let yourself rest when you know exactly where you are and how far it is to the trail end point.

We start counting our double steps. I use my fingers to keep count. I know that after I count 66 double steps, I have covered 100 yards and put up a finger. Another 66, double steps, raise the right thumb. Another, 66, my right forefinger is raised. We have traveled 200 yards. Another, 66, up goes my middle finger. Another, 66, I raise my ring finger, and after another 66, up goes my little finger, measuring off 500 yards altogether. Start over again using the left hand. When we reach the ruin, we take a few seconds to discuss the distances, directions and time passed and begin discussing the next navigation point. At the end of the night, we arrive at the truck and wolf down our breakfast before starting on the next navigation day trail. During breakfast, we doze off, chins resting on our chests until the order to start wakes us up. I fill my canteen with water, stuff some food in my army belt and off we go.

The second day's trail leads us from ruin to water hole and to a deserted orchard. At 10:30 a.m., we are sitting in the orchard eating oranges and spreading sardines on a piece of bread. We drink water. The sun is rising and we look for the next navigation point that will take us in a semi-circle for many miles to the end of this day's march. We reach the next navigation point and nap for a quarter of an hour. I wake up and jump to my feet. "We need to get moving." Lony and Yermi stand up and we hustle to the next navigation point. We are exhausted down to our bones. At the next point, a ruin within a sabras fence,* we discover that we have an extra 10 minutes and sleep. We wake up after the nap feeling almost refreshed and continue quickly toward the end point of this navigation day. There, we will eat and prepare for the coming night march. Two hundred yards before we reach the end point,

* Sabra is a prickly pear cactus and its fruit is sweet. The name has come to refer to any Jew born in Israel.

Lony sprains his ankle. He does not complain, but Yermi and I support him by placing his arms over our shoulders. In that way we reach the end of the trail.

The paramedic takes care of Lonny. The commander assesses the situation and consults with the paramedic. His verdict: This night only Yermi and I walk.

We drink warm chicken soup and study the night trail during the hour left before the night march starts. The truck driver looks at us with compassion as we fall asleep, heads nodding into our soup bowls. I go over the trail in my mind and it is simple. The navigation point before the finish is Tel Grar and after that it is a straight line towards the Beit Kama junction memorial, dedicated to the soldiers who died protecting the Negev during the 1948 Independence War. I memorize the trail marks and take another 10-minute nap.

I wake to the sound of the other navigation groups filling their canteens and setting off on their night trails. I calculate the time our walk is supposed to take and conclude that if we walk at a reasonable pace, we might have time for another hour of sleep before sunrise. The navigation towards Tel Grar is accurate and quick. I recognize the old Turkish railroad that connected Cairo and Damascus. We cross the muddy Nachal Grar wadi, walking between the reeds to the northern bank and see Tel Grar. Now we have a long trek in one direction to reach the end point of the trail. I say good-bye to the Mishmar Hanegev lights to the east and take the azimuth towards Beit Kama Junction. A thick fog engulfs us. I ask Yermi to walk ahead of me and hold my phosphoric compass in front of me and behind his back. It shows 83 degrees.

I tell Yermi, "When you deviate to the right, I will cluck. When you deviate to the left, I will tap on my rifle."

On we go. I can't see any farther than Yermi six feet ahead of me. I hold my compass up to my eye and see the phosphoric number and Yermy's shadow. I steer him to the right. We march. Adjust to and correct each deviation. We set a steady, fast pace. After about 5 miles, we enter a field of wheat and pea plants. The direction is clear and I barely need to use the compass, but we face a new challenge.

The combination of wheat and peas, covered with heavy dew, creates a thicket that makes walking difficult. To move forward, the leader must kick his feet forward to tear the bean shoots wrapped on the wheat stems. In the beginning, it is easy. I move step by step opening a path for Yermi, and he follows behind me on the open trail. With each additional step, moving forward becomes harder and harder. My leg muscles, exhausted after more than 40 hours of walking, are becoming weaker as I break trail through the thicket. I stop. Yermi stops.

"Yermi, it is your turn to go first," I say. "After that I will replace you."

Yermi leads. I hear him groaning. After 20 steps, he says, "Ouri, it is your turn."

We change positions. I walk in front and break trail. The wheat stalks are as tall as my navel and the beans wrap around my legs. After 50 steps, breathing heavily, I tell Yermi, "Your turn."

Yermi walks a few steps forward and we trade places again. Start walking with the knowledge that Yermi will stay behind me. Each step requires tearing the wet bean shoots. My thigh muscles are aching and tense. Yermi walks behind me. I lift my knee and slam my foot down. Lift the next knee and then foot down. The pace is terribly slow. The effort is unbearable. In the middle of the strenuous struggle to move forward, after our four exhausting, back-to-back trails, the commitment to make it to the monument seems far away now. A wave of frustration, anger and helplessness floods me and I begin to cry. Walk and cry. Walk and cry. I don't care about anything except getting to the end point and the waiting truck. Walk and cry. Uncontrolled tears.

How can I allow myself to be so weak? Mom moved from one line to the other. Always towards the line of survival and life. Auschwitz—Birkenau. Only after several weeks was she transferred to a factory for repairing airplane parts. During the weeks in Birkenau, she was constantly thirsty. "They barely gave us drinking water," she told me. She struggled, with stunning awareness, to stay alive and save herself from the murderous ovens.

My need to cry is stronger than the shame of crying in front of Yermi. To myself. I think that he is too in his own daze, can't see or hear me crying. I am ashamed anyway. The ingrained legacy of the survivor prohibits crying. Survivors like my mom who kept moving until they were set free by the Americans. Survivors like my dad who simply endured. No matter what. No matter the weather, with no food and little water. Endured. I'm mad at Lony for spraining his ankle. Mad at Yermi for not tearing the bean plants. Angry that I must exert such unbearable efforts. The rage bubbles up with my warm tears.

On I go.

With the same abruptness, my rage evaporates. Only the fire inside my thighs exists, reminding me at every step. My brain is paralyzed. How do I escape from this thicket prison? Right? Left? Each attempt to exit is met with frustration and disappointment.

When I was fifteen, the Six Day War erupted. The euphoria war. The intoxicating victory and the sense of unlimited power. After the war I read "Dialogue of Combat Soldiers" and other articles. I read a letter that a paratrooper wrote to his girlfriend about his desire to be strong and powerful so the atrocities of World War II could never happen again.

I felt a forceful connection to the need to be strong. My surviving Matkal—the Unit—stems in a large part from the desire to be a survivor—as a revenge, as a declaration that "I am alive and I will show you." The "you" is not defined but the other is the oppressor. The other is reminiscent of Tolkiens' heroes in *The Lord of the Rings*: Golem, the hermit who betrayed; Sauron, the evil behind the mountain ranges and his followers. All of them desire the ring to gain unlimited power. And I, the little Hobbit, fight stubbornly against the evil.

As a boy, I read *Undiscovered Affairs* many times. The book describes the English soldiers defeating the evil Germans that my mom hated so much. I would go to movies with her, and she would tell me, "Let's go see how the Brits screw the evil Germans."

The training in the Unit provided some temporary healing of my deepest fears and nightmares. Physically laboring until collapsing was a way to run away from my fears and avoid the unbearable feelings of helplessness, humiliation and oppression that my parents suffered, which I absorbed from them. The demons were held at bay for as long as I continued to do the exhausting walking of the Unit.

After a few more minutes of walking, I tell myself that there might be another way to get out of the thicket. A way off the azimuth line where the going is easier? I turn north towards Beit Kama. Walk a short distance and then return to the direction of the junction. I know there is a road nearby but don't know where is it relative to my position. Is it a shorter path to the road heading north or east? Since I don't know, I continue to walk northeast toward the junction. Marching through the thicket seems to go on forever. I walk at a slow steady pace and Yermi follows behind on the path I opened for him. For unknown reasons my body grows accustomed to this torturous walk. I have surrendered completely. I exert just enough effort to move forward. Maintain a steady pace that works for me. Pushing my way forward towards the junction is becoming monotonous. After one mile of threading my way through the thicket of wheat and bean plants, I emerge onto a dirt road. It stretches from south to north. I start walking north. The first steps are an intoxicating experience. I can walk again. Our legs are moving with amazing ease and our pace automatically quickens. My breathing slows and we walk faster.

We are getting closer to the junction. The truck waits there. We feel blessed to see its red tail lights. I climb into the truck and meet up with my other friends who took an alternative route and avoided the bean / wheat thicket. I continue to digest my crying experience. I do not tell anyone. I am shaking from the combination of the excruciating effort and my inner breakdown. Everyone drinks water and finds a corner of the truck for a short nap.

There is little respite because the ride to the sand dunes near Rishon Lezion for the last navigation of this drill takes only an hour. This trail is only 10 miles long and will mark the completion of the azimuth navigations. After almost 50 hours of continuous walking, what is another 10 miles? A trifle.

I barely remember that walk. We climbed up the Rishon Lezion dunes and rolled down. I felt like a blowing leaf, gliding with the wind towards the end point. The sun warmed and caressed us. At times we paused, took a few sips of water. During these breaks we wallowed in the dunes. The sand grains joined the dust and sweat accumulated on my skin over the past several days. When sand filled my boots, I took them off, dumped them out and enjoyed the sensation of my sore feet resting on the warm sand. Put my socks back on, shake sand out of my boots one more time, put them on, tie my boot laces tightly, and place the pant leg over the boot, securing it in place with a rubber band. This hopefully will keep sand from accumulating inside my boot. Yes. Less sand, but some…still gets through. Remember the Little Prince lost in the Sahara dunes. I do not want a rose. Grow accustomed to the soft sand and being embraced by the dunes. Softly sinking into the dunes that welcome my body and become softer during this final day of this exhausting challenge.

I walk at a regulated pace. My footsteps became softer as well and move to the wave rhythm of the dunes. Slow down as I climb and speed up, just a bit, as I descend a dune. Up again and then down. The small amount of energy this requires doesn't allow for a faster pace. The soft warmth of the sun soothes my tired muscles and blends with the sand. Do not care that my machine gun is loaded with sand. That grains of sand adhere to every sweaty cell on my skin. Want to arrive but not to hurry. Feel a mix of 'Baduizm, The Little Prince and tiredness that encompass us all. Walking feels weightless, adjusted to the air, warmth, light and sand as I "glide" northward. Go up and down the dunes, enjoying the caressing sun and the soft sand. Recall the hard touch of the soles of my feet on the loess during the last two long days. As I ascend the soft sand my pulse rises. When I descend, my pulse slows. Early afternoon we reached the truck parked on the road north of the dunes. We climb in and fall asleep almost immediately after nearly 60 straight hours of walking.

Arrested by Military Police

Judges and officers shalt thou make thee in all thy gates.
(Deuteronomy Ch. 16, v. 18)

A month after completing training and covering thousands of navigation miles I became ill. Because I had a high fever and felt very weak, the doctor recommended that I rest a few days. The commander suggested I go home to recuperate while my crew members remain in the base and continue further training. I took him at his word and left the base wearing my field uniform and with no written leave permit. I did not look in the mirror, but I must have looked pretty awful, unshaven in wrinkled dirty clothing with bloodshot eyes and a 104-degree fever. I hitchhiked with drivers who stopped for me and reached Kastina Junction quickly.

Walking toward the deserted road that headed south, I hoped to quickly arrive home to my bed and mom's treats. As I hailed a car, all of a sudden, a military police jeep pulled up next to me. Three MPs approached me.

"Soldier, do you have a written leave permit?"

"No…I am on my way home," I stuttered.

"Show us an ID."

As expected, I had no ID on me. I was nervous, recalling stories about soldiers who ran from the military police. My anxiety got the better of me, and despite the fever and weakness, I started to run, clumsily, as if 500-kilogram weights were attached to my legs. The MPs, with no real effort, caught me quickly and pinned my arms. They dragged me to the police jeep and drove to their nearby base. There I was, despite the months of rugged training, easily caught and in serious trouble. My limbs felt feeble and my heart sank.

I asked for water. One of the policemen gave me his canteen and I sipped. The jeep sped down the road. I began to collect my thoughts and decided to tell them that I am a soldier from the Unit and that they should contact my base. I was worried they would not believe me, considering my frail appearance and sloppy uniform, but they called.

The message reached Ehud Barak, and he requested that I be brought back immediately to the Unit. Before long, I stood facing him, with grave concerns about my fate.

I heard Ehud say, "I sentence you to one month detention on the base. The reason is that the policemen could have shot you when you ran instead of talking to them."

In shock. A truth I did not want to confront.

I walked back to my room and collapsed on the military cot, far from home and away even longer by a month. I felt humiliated, tired and even sicker than before.

Barking Dogs Bite?

Deliver my soul from the sword; mine only one from the power of the dog.
(Psalms, Ch. 22, v. 21)

With each additional night navigating in the desert I came to know the night better. Darkness became my friend and I enjoyed its embrace. I felt protected and secure. One time we went, Yadi and I, on a couple of night navigations in the Negev. We walked through a cloudless, star-filled desert night, smelling camp fire smoke and listening to the distant sounds of barking dogs carried by the light wind. A bird spreads its wings, rises, and disappears into the darkness. The sound of crickets. The smell of thick shrubs that line the sandy, flat wadi bed. Through my boots I feel the changing texture of the creek bed transitioning to a hill slope, covered with sharp flint or limestone.

I think about the Negev rocks eroding at a time table vastly different from mine, whose entire life passes in the blink of an eye. Negev Geological time. By comparison, we are arrogant ignorant nomads quickly passing by.

Traveling eastward. The terrain becomes rugged and furrowed by deep canyons. The navigation trail is clear and we hurry forward. The night is half over. We stop for a moment to drink and discuss the trail ahead. The wind cools me and dries the sweat that has accumulated on my military shirt. We get up and resume the familiar, brisk marching speed.

The smell of camp fire smoke and the sounds of dogs barking, louder now, are carried by the wind. The sounds of human voices join the mix. Climbing uphill. Our breath shortens and our pulse rates rise. Looking up at countless stars in the clear night sky, we feel like a veteran navigation team existing in a walking reality. Hearing familiar sounds and smelling familiar smells, within a comfortable darkness that is calming and pleasant deep in the night.

The fire smell becomes stronger and fills my nostrils. Walking uphill. Out of the darkness, the shape of a Bedouin tent appears. A pack of barking dogs approaches us. Fear rises from my belly. My sense of calm disappears. I lean forward and collect a few stones in my hand. The trail passes a few dozen yards to the right of the tent. We continue and the dogs are coming closer. I throw a stone at the leading dog and growl at him. It dodges the missile and retreats. Yadi giggles. The pack barks even louder and my tension rises further. The pack keeps back at a safe distance but continues to follow us. Yadi and I march at the same pace. I turn and look back at the trailing pack. Feel the tension and get ready for their attack. One stone held between my thumb and forefinger, ready to hurl at them. Two more stones held between my ring finger and palm. Ready for them to charge us. Gradually moving further away from the tent. The sound of their barking is fading in the distance. I let the stones drop from my hands. My breathing relaxes as we descend into the wadi. Look back again. The night is back to normal. The sound of barking fades away.

The trail turns north. One more steep wadi and we will head west by northwest. The wind fades and our tiredness rises. I feel the familiar pains from overexertion in my legs. Recognize again the soreness of muscles in my thighs, calves and feet, and pain in my right shoulder from carrying my Uzi. Each muscle waits its turn to report its specific pain. From time to time, my boot slips on a stone that slides with a unique "kehhhhh" sound. A quick recovery step and we are back to our customary marching pace.

Beginning of third watch. We are walking up a mountain branch that leads to a plateau. Enter a barley field. The grain shells crunch under the crepe soles of our boots. The smell reminds me of years working in the barley and wheat fields of Kibbutz Mishmar Hanegev. It is a familiar and beloved smell, and soon replaced by my joyful thought, *In an hour, the sun will start to rise.* We cross the threshing circle and head west.

Suddenly, out of nowhere, a huge barking dog leaps towards me. Fear floods my body and I am instantly fully awake. I start running, the dog following close behind.

Yadi shouts, "Stop."

The dog is coming closer. My fear intensifies by the second. Ready myself for the dog's attack. Remember I am carrying my Uzi. I turn around pointing the Uzi at the dog, ready to shoot it. Both hands on the Uzi, gun butt at my belly. Bend my knees a few inches to steady my position. My fear transforms to combat readiness. The dog stops, still barking ferociously. I stand still, finger on the trigger. I breathe steadily while we stare at each other. Yadi throws a stone at it and, at a blink, the dog begins to whimper, turns and runs away. I continue to slow my breathing as Yadi asks me, "What happened? Why did you get so scared? Don't you know that you should not turn your back to the dog?"

My legs and hands start shaking. I take the racked cartridge out, ensuring that the Uzi's barrel is empty, and put the cartridge back. We continue walking. After two hundred yards, we re-check our position on the trail. I am still shaking from the intensity of my fear of the dog.

Walking again. The presence of Yadi and the steady marching pace calms me. I console myself by walking and gradually become calmer.

I have always been afraid of dogs. When I was twelve, I returned one day from milking the sheep. It was dark and I day-dreamed as I walked, passing by the milking shed for the cows. Suddenly, from a parallel path on a small rise above me, a dog leapt towards me. I was panic-stricken and ran as fast as I could, certain that the dog was right behind me. I ran the 400 yards at record speed. I don't think I had ever run that fast in my life before.

I must have absorbed the fear of dogs from my Mom. To this day I can still recall with dread her stories about the Nazi guard dogs that were trained to jump on prisoners and tear out their throats. I was afraid of the kibbutz dogs as if they were the same guard dogs that killed Jewish prisoners. Trained to take down inmates in the camp, or during the daily walk to work, and on the final death march. All those images and the fear that accompanies them are etched in my mind and still shake me. I feel the rage stemming from helplessness. I can see the German guards pointing at a prisoner in striped pajamas. The dog leaps. The prisoner raises his arms in an attempt to protect his face. The vicious, well-trained dog takes the prisoner down and savages him. Remembering, I almost jump out of my skin. I'm trembling while I march.

How did my mind travel to the brutal cold of a Polish winter? I am here in Israel, on a clear night, walking in the familiar terrain of the Negev. What am I searching for in the barren landscape of eastern Europe's winter?

It is so good to be walking, to feel the comfortable warmth of the desert. I suffer the aches of prolonged marching, but what are these pains compared to the helplessness and humiliation of my relatives in Europe in the 1940s?

I have already learned to wrap myself in the warm embrace of the invigorating night. To feel the close secure embrace of the night's familiar darkness.

We continue to walk. Traverse the last azimuth and hasten our steps towards the end point of the trail. Familiar first light of dawn emerges behind us. Our tiredness diminishes with the knowledge that we are nearing the waiting truck. Feelings of gratitude imbue my senses as the light gradually fills the Negev morning skies.

Gentle winds caress us. We run the last hundred yards, arriving at the truck breathing heavily, and smile at each other. We did it again. We load our military belt on the truck and take our positions on the mattresses inside.

Terrace Crossing

After our time in the Galilee, we navigated the West Bank, which has many terraces. They were built over many generations by farmers who planted trees and crops and fertilized the hilly soil. The terraces are made of stones and create walls that rise from a foot-and-a-half to over nine feet, depending on the steepness of the slope. On the flat areas between terraces, the farmers grow olive trees, various fruits, or seasonal crops. Some of the terraces were constructed so beautifully that one can imagine the elevation contours on the navigation maps drawn based on them.

As a navigator who must traverse terrace after terrace all night long, the challenge is to make the crossings with minimal exertion. We devoted much effort to developing a terrace-crossing protocol for our navigation trails. Each terrace has paths that enable farmers to move up and down its field. It is preferable to walk on these paths, but they were not designed with navigation in mind. In most cases, the trail did not conform to the terrace path.

Some terraces were so high that I couldn't reach my hands over the edge to pull myself up. In that case, I would walk along the terrace until I reached a point where I could climb up and cross. Descending was usually easier. No matter where I stood on the terrace, it was possible to climb down. The fastest way was to jump down, landing on my feet and continuing to walk. Another option was to lean on my right hand, walk half a circle, jump and land. If the distance was too great, I'd slide down on my butt, taking into consideration that my pants could tear. Small tears were not noticeable, but some larger one left us barely covered, as if we'd been dressed by some top-tier innovative fashion designer. Some of us became ground-breaking fashion designers ourselves, down to only one pant leg by the end of a long navigation trail or pant legs that were slit all the way from top to bottom.

We were concerned only with moving past the terraces quickly and reaching the trail end. We concentrated on balancing speed and navigation accuracy with minimal energy exertion. The preferred trail was along the farmers' paths, but many times I had to cross where no path existed. During winter, mud became a huge factor. Walking the farmers' path ensured that lots of mud would stick to our boots and accumulate on our soles. Each new layer of mud added to the prior layer, creating a build-up. I found myself lifting huge, heavy limbs weighted down by layers of mud so thick that they hid my boots. In order to forgo the wonderful delight of carrying mud from one field to the next, I would scrape my boots on the terrace stones. For once, we were glad to have these stones around. Usually, they caused us to stumble and interrupted our progress.

The style of terrace crossing differed from soldier to soldier. Shay would bulldoze through the terraces, creating an additional passage and easy descent for other crew members. Dany would jump to the upper edge of the terrace, lift himself up and move on. His physical strength lasted for many miles. It was only after we traveled long distances that he reverted to a more "conventional" terrace crossing style. Amnon would place the tripod of his RPD, a Russian machine gun, on the edge of the terrace, secure it, and pull himself up. Yehuda's soundless laughter would accompany this magic performance. I would place my knee on the edge of the terrace, lean on my palms and pull myself up. At some point, we all tried almost everyone else's crossing style until we found the style most suitable for ourselves.

Most terrace crossing styles created additional passages in the agricultural fields. Other night navigators could take advantage of passages we made just as we profited from the trail blazing efforts of our predecessors. No doubt, we created additional work for the farmers who woke up to new passages cut through their fields. At least we didn't cause too much damage.

The Munich Olympics Massacre

*Ye have multiplied your slain in this city, and ye
have filled the streets thereof with the slain.*

(Ezekiel, Ch. 11, v. 6)

OUR ATHLETES KIDNAPPED IN MUNICH!

Terrorists infiltrated the Olympic village and kidnapped the Israeli athletes.

Right away we in the Unit start to pack and organize, preparing for a rescue mission. We are being deployed to Munich to save our athletes.

Where is it? We get a map of Europe and find Munich.

We have our pictures taken and, a few hours later, receive our passports. Gathering in the briefing hall, we get the first detailed information about the situation. Assemble our military equipment. Our government ministers are negotiating with the Germans to allow us to lead the rescue mission that will confront the terrorists. Yadi and Adi reset the snipers' telescopes on their rifles. We are issued light army belts. Face to face combat is expected. Knowing that the fighting will be at close quarters, I replace the Kalashnikov with my familiar Uzi. We hone our skills at the shooting range. Yehuda takes a picture of us by the climbing ropes and I take one of him. Another briefing. We arrive at the Ben Gurion International Airport with our military equipment. As we wait for the plane to take off, our excitement ebbs. We sit, munch on snacks and doze off. I start daydreaming.

My Dad's eldest sister was Shiendele. She was a skinny girl who helped at home and worked in the small store of my grandparents. She was a brilliant student. Never made a mistake on her exams. Despite the efforts of the teachers in Krakow to fail her, she received highest honors every year and was the best student in her school. A Jew. In Poland during the years when Jews were persecuted and humiliated. The Poles were mad at her and jealous. How dare she? And Shiendele? She probably did not "dare" but couldn't help but be brilliant. Up to the last day at school when Jews were allowed to study in Krakow. The school gates were closed for Jews, who were forbidden to study.

Then the family moved to the Krakow ghetto, to Tragova 5. The house near the Pankievich Pharmacy. *Aktionen.** That was the last time my Dad saw her. He carried the memory of Shiendele with him through all the endless, nightmarish days of the death and labor camps. During the final death march when so many perished. In the ship carrying camp survivors from Europe to Israel.

As we were growing up, he would tell us of Shiendele, the best student in Dad's Krakow school. The story simmers inside me. I'm filled with rage and a desire to avenge her terrible and needless loss. The other siblings of Dad were also murdered. Their names are engraved on the memorial wall of Beit Borchov in our kibbutz. Sheindele, Israel Zeev, Sara-Ita, Zvi, Rosa, Yehudit (the youngest), father Tuvia, and mother Miryam Ahuva. My mother's family suffered a similar fate. She grew up in Geur, a town in western Hungary. Her brother Haim died of pneumonia at a young age and she was her parent's only daughter. All of her family was murdered in the ghetto and death camps.

* *Aktionen* refers to the raids of the ghetto few times a week when the Germans "collected" any Jews who were in sight. They brought them to the plaza near the Pankievich Pharmacy and loaded them on trains to be sent to the extermination camps.

Dad and Mom were the only survivors of their families. There were many brilliant students in their families, but no fighters. Only later did I start to understand what kind of resilience was required to survive the European inferno through the many years of the Holocaust.

Mom does not talk about it much. She takes great care of us and worries about us. Tells us a little here and there. Told us how she arrived at Auschwitz with her grandmother. Was elated when the war was over and spent happy days with her orphan friends in Blankenese, a Hamburg suburb. But I know very little about Mom's experiences during the war. Most of what I know is foggy and fragmented.

When we ask, Dad tells us details. As we hear more stories, we come to understand that he was probably "chosen" to survive by his father. Survive and build a new life and a family. As his eldest child, I felt that Dad expected more from me. All the time demanding more. No words were uttered but that was my implicit understanding. Mom took care raising us, with great anxiety that something bad would happen to me. I felt a duty to excel but at the same time sensed that, no matter what I did, I would never be able to meet the unexpressed expectations that were neither defined nor described. I would always be missing something, somehow defective. Living in a state of anticipated and unavoidable failure.

Auschwitz, Buchenwald, Birkenau are all names of places that evoke rage and dread. When I hear these names, I'm flooded with feelings. Thoughts of the Nazis' murderous machine that annihilated members of my family evoke fear and terror. As a boy, it was hard for me to relate to the Holocaust background of my family. The feelings were blocked, avoided, unseen. I was born with dead uncles and aunts. Murdered. Besides Dad, Mom and myself, dead relatives were always close by. I used to look with burning envy at my class mates who had uncles, aunts, grandmothers and grandfathers. I felt alone and isolated. On the Shoah Remembrance Day, when the six torches were lit, I cried but hid my tears, leaving the ceremony to hole up in my room. During vacations, when other children went to visit their relatives, I went to my parents' friends. During these visits I felt different, like a stranger. Someone who never would fit in. The worried looks from Mom and the expectant looks from Dad were constant companions that weighed me down.

We were sitting on the military combat equipment at Ben Gurion Airport waiting for the German government to give permission for an Israeli commando unit to go to Munich. The Germans don't want us to lead the assault because they "know" better. They are convinced they can handle the situation and free our athletes. Convinced that they "know." Always they "know." They are orderly, organized. But, they do not have what we have. We are children whose parents saw both lines in the concentration camps.* And jumped from one line to the other to survive. And survived. The cunning determination of the survivor is our legacy from parents who were forced to choose lines. They never felt the anxiety of inevitable death. Our training integrated with their instinct to survive. The Germans did not experience dread the way our parents did. They were comfortable, on the safe side of the fence.

Sitting. Brooding with a sense of urgency and concern about what might lie ahead. In the airport. Waiting. Hours pass. Final decision: The Germans do not give permission for us to come and lead the operation. None of our leaders can convince them that we are the best, most qualified to lead the mission. It seems that there was never a chance to convince those who were on the German side of the fence.

* Upon arrival in Auschwitz, prisoners were divided into two lines—death and forced labor/life.

We fold our gear and return to the Unit. Later, the radio announces the dismal failure of the German operation to neutralize the terrorists in Munich and protect our Olympic competitors. And the price. Eleven of our athletes paid with their lives.

Jeeps Are Soldiers' Best Friends

Behold, he cometh up as clouds, and his chariots are
as the whirlwind; his horses are swifter than eagles.

(Jeremiah, Ch. 4, v. 13)

The commander asked me to look into the possibility of having the crew sleep in Mishmar Hanegev for the first two nights of Jeep navigation training. That got us excited and joyful before the Shabat Holiday, anticipating a week of navigation with jeeps in the south.

During the many weeks prior to the jeep navigation exercises, we labored long and hard. We began each week after a restful Shabat with another round of arduous, exhausting, sweaty navigation marches. I experienced muscle aches, blisters, bruises, hunger, thirst, labored breathing, heart pounding, and other seemingly endless challenges. Continued physical and mental extreme exertion. Our sweat, mixed with dust and salt, accumulated on our skin during the long week of marathon marches. Rain soaked our sweat-covered bodies and caked our army uniform, pants, shirt, underwear and socks with mud. The mud leached into our skin. Mud and sand permeated my being and created a crust of indifference regarding my appearance. I felt like an ox plowing an endless furrow with utmost efforts. And then another, and another.

We continued to drive ourselves: "Get to the land mark. Move it! Finish the trail."

When we arrived at the base on Sunday, we knew we are going to have fun in the south. Not just have fun. A week of jeep navigation was the realization of our dreams. To cover a piece of land not on foot, was the most exciting thing we could have imagined. It is hard to describe the joy we experienced as we received the long equipment list and saw at the head of the list the first item staring at us—JEEP.

We are four teams. As we equipped our jeep, we became even happier, knowing that it would transport all of the heavy equipment we usually had to carry. We are not going to carry even one gram of extra weight on our backs. We allowed ourselves to think more freely and added extra items such as mattresses and sweets we brought from home or bought at the local IDF grocery. The jeep carries everything. As we finished loading the jeep, there were maps of the south, clothing, blankets and enough food for an extended "holiday." The heap of equipment loaded on the jeep is the striking difference between a northern, walking soldier and a southern, jeep-riding soldier. The foot soldier is tied to the trail attempting to complete it on time and with some energy in reserve. The jeep-riding soldier has the freedom to look at the scenery. The pleasure of navigating in the Negev by jeep became even greater when we passed a trail we had recently walked. We visited each spring and cliff marked on the map that was some distance from the trail and had piqued our interest. Physical limitations disappeared as a slight push on the gas pedal carried us a mile or so down the road. We felt elated by the expanse of the desert. We updated the familiar adage "The route is wiser than the walker," by adding "but the Jeep is even wiser." It's good to be the jeep navigator, but even better to be the driver. You appreciate every moment of the trip, feeling the vast difference between pressing on the gas pedal and putting one aching foot in front of another during an endless, exhausting march.

Walking soldiers count distances with yards of sweat and muscle aches. Covering distances with a jeep creates a sense of elation and frees much time and energy for thinking. Most of the physical energy that would be consumed by marching remains available. The constant struggle between mind and body does not exist. The desire to give up never even crosses my mind.

With our four jeeps fully loaded, we sat down for a briefing with commander Dany Dagan, our mechanic. Dany could be our father. His long-time civilian experience as a theater producer and business man made him our ideal spiritual guide. I learned a few key sayings about driving from Dany such as, "If there is a doubt, there is no doubt." This was useful advice for the situation just before overtaking the car ahead of you on a one lane road. Another one was: "Average drivers try their brakes at the curve. Experienced drivers brake before the curve. Professional drivers test the brakes at home." Equipped with these initial driving lessons and expecting more to come on the road, we embarked on our first week of jeep navigation. To miss as few hours as possible, we ate an early lunch, gladly foregoing Sunday physical training, and off we drove.

At the navigation starting point, we were three men in each jeep. Driver, navigator and passenger. The soldier riding in the back earned the nick name "Chimidan" (army bag)—since his only task was to be carried. Zero responsibility. We took the "Chimidan" concept with us for many years. After completion of active duty and into the reserve days, you could still hear from former Unit members say, "I am so tired. Do you mind me being a Chimidan for the next hours?"

We were handed the day trail maps and shown the gathering point to prepare for the night ride. At the beginning, the jeeps rode close to each other. We hailed our friends in the open vehicles joyfully. Our shouts to each other diminished when our commander rose to his feet. We enjoyed everything about riding in the jeep: the roar of the engine, the grinding of the gears, the unique sound of the tires riding on the loess ruts, the vast wheat fields and the open skies. Enjoyed smelling the Negev scents carried on the blowing wind.

I experienced the familiar sights and smells from my days and nights walking the kibbutz's fields with my friends Lotan, Meir and Yishai, escorted by the irrigation manager Zvoki. As children, we would go to the tractor mechanic shop, wait for the field jeep and hope Zvoki would agree to take us with him out to the fields. We felt a great sense of freedom traveling through the open fields of the kibbutz.

As we approached the end of this day's trail, we suddenly hear shooting from a close distance. On our right, a field of tall wheat ready to be harvested. We look in the direction of the shots and see Dany Dagan's jeep enter the ripe wheat field and accelerate. A few more shots and then excited shouts from Dany's jeep. All of the jeeps arrive near the edge of the field next to Dany's jeep. We see a gazelle dead on the ground.

Dany Senesh says, "Why did you kill it?"

One of us replies, "What's the matter?"

Dany accuses, "You are Hooligans! Just to kill?"

Silence.

Within a few minutes the gazelle is tied on the hood of the jeep and the navigation journey continues. I can smell the fresh scent of the broken stalks of wheat. At the end of the day's travel before we begin the night navigation, we talk about our experiences, some about the navigation, much more about the gazelle meat we expect to eat later tonight.

We begin the night navigation route. The day's excitement is over and we're tired, forcing us to focus on navigating and driving. The backseat passenger appreciates being a "Chimidan." At times we contact the other jeeps to make sure all are on-track to reach Mishmar Hanegev before dawn. The kibbutz night guards already know that we're coming. Around 3 a.m. we arrive at the gate and call for the guard on the intercom. Uncharacteristically, it takes him all of a minute to get to the gate. Three of the four jeeps enter and park near the old dining room, which also serves as the gym hall. We bring our sleeping bags inside and spread them on the floor. Some of us, including the commander, crawl into the sleeping bags.

I suddenly realize that Dany Dagan's jeep is missing. "Where are they?" I ask.

Nobody knows.

"Who spoke with them last on the radio?"

Apparently it happened a while ago. I attempt to contact them on the two-way radio. No response. Go outside to the jeep and try with the stronger radio. No response. I grow concerned. Where are they?

I go to the commander, who is already snoring, wake him up and report that Dany's jeep isn't here and we can't raise him on the radio. He mumbles something. I tell him that I am going out to look for them. He mutters "OK" and goes back to sleep. Strange. He is the commander.

How can he allow himself to go back to sleep?

During many of the Friday briefings before we went on the Shabbat vacation, he stressed how important it was that we be able to function when it is difficult. *What about the moment of truth? How can he continue to sleep? Not concerned about the missing crew members?*

I find Ovad who immediately joins me and we get in our jeep and head towards the trail Dany and his team were supposed to take. We drive around and attempt radio contact every few minutes. No response. After an hour we return to the kibbutz. My concern is growing. I report to the commander that we didn't find them and that I will continue to look for them. He mumbles, "OK," changes position, goes back to sleep and is soon snoring.

As soldiers we take responsibility. There are no discipline issues in the Unit because every soldier assumes responsibility as if he were the commander. The need to be accountable is ingrained from the first day of training. One of the lessons driven home to us time and again, especially during the operational stage, was "In the field, you are the chief of staff." No one else but you. Being responsible for our success or failure in the field is a heavy burden. Our only focus is on performing the mission. Each crew member is responsible for making sure that the mission will be completed perfectly. The work requires seriousness and focus. The seriousness affects other parts of life. Freedom exists only within a rigid framework.

Ovad and I go out again, deciding to widen the search perimeter. Both of us grow very tired as dawn breaks eastward along the tight loess roads. We decide to head further east to the point we passed at 1 a.m. last night. The rising sun blinds us. We try over and over to contact them on the radio with no success. Suddenly, as we speed east, behind a terrain fold, barely a yard from the road, we see a camp fire. Around the camp fire we see our friends and the jeep is nearby. Each one holds a stick with a piece of meat on it, roasting over the fire.

We call excitedly, "What happened? Why were you hiding? Why didn't you respond to our radio transmissions?"

"Come. Sit down," Dany replies calmly.

In a second, happiness that we have found them replaces our concern.

"Everyone was worried about you," I say.

"Stop worrying. They are all asleep," Dany says, grinning. "Come and eat some of this wonderful meat."

We hesitate for a second, then sigh with relief and join the party. As we chew the tasty gazelle meat, we ask, "What happened to you?"

"Our jeep broke down. We waved to the jeep that passed us, but that crew waved back and didn't stop. Since we couldn't get the engine to start, we turned off the radio and decided to cook and eat the gazelle. We figured somebody would come for us eventually. We were hoping it would take longer." Dany chuckles before adding, "Come and have some nice hot tea."

Why did they turn off the radio? The question perplexes me, but I let it go. Sitting on my butt. The tea is heating on the fire. Look out over the beautiful wheat field. At the edges, yellow flowers grow and white butterflies hover calmly and fly lightly around. The sun rises lazily.

Drink tea, eat some of the succulent gazelle meat. How tasty can such meat be? Heavenly. The taste of that meal is carved in my memory as the best ever. Divine taste that can never be experienced again. I enjoy each moment of this Negev morning.

Dany Senesh

Then even he that is valiant, whose heart is as the heart of a lion.
(Samuel B, Ch. 17, V. 10)

During the last night of another jeep navigation in the south, as we hailed the wisdom of the Jeep, we chose who would take what shift standing guard and went to sleep. There was a misunderstanding between Doron and Dany, and Dany was caught sleeping during what should have been his guard shift. When we returned to the Unit after soaking up the beauties of the desert, we learned his punishment. Dany was restricted to the base for the weekend and required to attend the flag lowering ceremony every day for a week. At that point, we were already an operational crew. We all were sure that such punishment was excessive since the flag lowering ceremony was performed by the newbie soldiers who had just arrived at the Unit.

Dany took it as a deep insult and was unwilling to go along with the punishment. I offered to stay with him over the weekend and for every flag lowering ceremony. The other crew members tried to persuade him, too. Nothing worked. Dany said he was unwilling to stay with our commander. His attitude was: "How dare they to punish me in such a way when I am an operational soldier? I won't put up with it!"

Because of the extreme demands put on us every day, it was hard to see the point of adhering to strict discipline standards. The Unit commander, Giora Zorea, and Amitay Nachmani, the Unit operational commander, felt that a punishment was a necessity. The impasse resulted in Dany leaving the Unit and joining Sayeret Golani, another elite unit, where they welcomed him warmly.

Dany Senesh had red hair and boundless energy. Every Sunday, Dany and Ovad were the first to cross the finish line of the one-mile race with our commander. Dany was straight as a ruler and had high degree of self-respect. He always completed his assigned tasks correctly, completely and without hesitation. He operated based on his established principles and wholeheartedly supported his friends. Volunteered for every mission without questions. In our crew, he belonged to the "right" politically and supported Yadi and Adi in their arguments.

Dany was an impressive person. Yet, despite his outstanding athletic abilities, he stayed humble and regarded himself with a sense of slight self-mockery. His perspective of us and himself was: "Do not think too highly of yourselves. There are people who are so much better than you." Frequently, when we became too arrogant, he would be the one to deflate our balloon of excessive self-importance. I knew that if I ever needed assistance, Dany would be there for me with no hesitation and no questioning of what he would receive in return.

Dany left us and joined Sayeret Golani. He was killed storming a house in Naharia where terrorists were holding some civilians hostage. He led the assault and was killed while attempting to free the hostages.

Dany, my dear brother,

It has been over thirty years and I still miss you. I wake up at nights and your memory is still alive inside of me. Miss you.

Today, so much time has passed since we were together, yet you are still present inside me. I remember your smile, your way of speaking, your piercing insights into reality and truth. I miss you as if you had left yesterday and were traveling thousands of miles away from us.

I remember that we tried. We were all standing on the asphalt trying to convince you to stay with us. We knew that you would act on your principles. Walk away from us. And probably no touch or words could convince you to stay. You'd had it.

My brother Dany, I miss you. You left a mark on me. I remember your love of life and your commitment to your high standards. You lived among us with sweeping passion and energy, both in the IDF and at home in Moshav Ben Ami.

You always responded to injustice with courage and treated others with respect. I remember your sharp sense of humor during the difficult hours, laced with self-deprecating irony.

Since then and thanks to you I know where to look.

I hope that when I arrive in the places where you are now, we can reconnect.

Rumi, the 13th century Persian poet, died in 1273. On his tomb is written: "After I die, do not search for my tomb in the earth, search for it in the hearts of my people."

I want to thank you Dany. I was blessed with the wonderful opportunity to be with you and to carry you in my heart.

Love you.

Your brother, Ouri

Wrestling—Krav Maga

And Jacob was left alone; and there wrestled a man
with him until the breaking of dawn.

(Genesis, Ch. 32, V. 25)

We practiced wrestling. It suited our juvenile fantasies. I thought that after we had learned the martial arts, we would overpower enemy soldiers guarding well-hidden places. Then we would penetrate the concealed locations and steal secret documents, maybe even save an Israeli spy like Eli Cohen.* And come back, welcomed like heroes but...of course, still humble.

In the meantime, while the IDF chief of staff didn't yet recognized our capabilities, we learned how to knock each other down on the sand, how to fall, how to free ourselves from a rival's hold and attack his vulnerable points. At the beginning of the training, when we hit our friend's vulnerable spots, we did it carefully and softly. The wrestling instructor saw this and attempted to goad us to attack more aggressively. He had little success until we started fighting in earnest.

During the matches our competitive natures took over, and we actually fought. Victory required overpowering a rival and pinning him to the ground. As each pair fought, the rest of us surrounded them in a circle, watching. We all fought. One time, after pairing up

---

* An Israeli spy who infiltrated the Syrian military and government by posing as a businessman from 1961 to 1965. Eli Cohen was hanged in Damascus and his corpse was never given back to Israel.

Avner had no partner. Anyone of us could have wrestled Avner, but unfortunately for him, the commander chose to be his partner.

Avner fought better and overpowered the commander. As he was sitting on the commander, enjoying his victory, he gave us the thumbs up sign, smiling broadly. We were all elated, but the commander noticed him gloating and was deeply offended. We didn't know it at the time, but at that moment, Avner's fate in the Unit was sealed.

Each crew in the Unit is named after its commander, who builds its qualities and capabilities and helps mold its character. The choices and operational successes or failures of the crew originate from the way it is trained. The combination of each soldier's desire to succeed and excel and the high standards set by the commander, shape the environment and determine the character of the crews.

The job of the commander is largely unappreciated. Because of the extreme demands made on the crew, the commander can't be a friend to his soldiers. During the intense field training, he is alone. His only potential friend is the escorting driver. The crew as a group unites against its commander. In many combat units, there are two officers or a sergeant and an officer. Having a lone commander leads to faulty decisions, mainly in the area of human relations. Many good, competent people fail as crew commanders. The relentless operational demands and the uncompromising requirement to complete the mission are accomplished at the expense of human relationships. Most Unit crew members are opinionated and convinced that "The mission could be achieved in another way." While performing demanding missions, much tension develops between the commander and his crew members. The hardships of the missions lead to significant suffering. Yet, from these sufferings a cohesive crew is born.

The commander makes continuous, unrelenting demands on his combat crew. He is the representative of all "unattainable" capabilities. We were afraid of ours. The demand for high standards and attention to detail, a part of the Unit tradition, were inculcated in us. When we started training in the Unit, none of us knew how to operate according to these standards. We considered the commander as the ultimate representative of these professional standards and capabilities. The huge gap between his proficiency and ours was apparent. Axiomatic. Intense and demanding training lay ahead for us, and we knew that we would be expected to accomplish "impossible" tasks. At the same time, we fear failure and dismissal. All of us embraced the training with full commitment. As we continued our training in the Unit, we saw that what was required from our crew was more than the other crews and we developed team pride. During our training we heard that we were regarded as a "strong" crew. Yet, I could see that good soldiers were dismissed for no clear reason, and I dreaded the possibility that I would become one of them.

During the initial weeks period, the commander led the excruciating trainings. After many months of trainings, we noticed gaps between the demands he made on us and his ability to meet those demands. During the intensive training period, each one of us built very high mental and physical capabilities, to the point that our commander had difficulty keeping up with us. As we became an operational crew, the gaps, especially in areas of endurance and perseverance during strenuous challenges, widened. And with them, our trust in our commander declined. We no longer trusted him and the crew's relationship with him deteriorated.

Through the years I have repeatedly asked myself: How did we become a highly skilled and capable crew while having such rotten relationship with our commander? How does the commander—crew relationship affect operational capability? These questions continue to nag at me.

Crew commanders in the Unit had unusual power. As the commander is chosen to command, he is given unreserved trust. In most cases, the choices are good. The training platoon commander gives him absolute authority to train his soldiers in any way he sees fit.

The commander chooses how much to press and how much strain to put on his soldiers. There is a tradition of setting high standards for the length of navigation trails and the hours of fitness training. The crew commander can add a few more miles to each navigation trail or extend the fitness training by a few minutes. After many years of training, human limitations are known and the physical challenges can be modestly extended.

In a system where the crew commander has almost complete freedom and total trust, a crew member has practically no options if he is unhappy with a command decision. If the commander likes a soldier, he has a good chance of completing the training period; but if the commander dislikes a soldier, even someone with very high capabilities, the odds of him making it through training are very low.

There was no doubt that Avner was an above average soldier. But as soon as he embarrassed the commander, his path to dismissal was a foregone conclusion. Shortly after the match, we heard from the commander that Avner was 'below standard' and didn't make a 100% effort. In reality, nothing about Avner had changed, but it wasn't long before he was dismissed from the Unit. At the time, we did not talk about it among ourselves. The "monastic silence" was integral to our existence in the Unit. We did not spend time analyzing the wrestling incident and its results, yet for many years after, questions about it haunted me. Years later, I found out that my friends were also troubled by these questions.

Maybe we should have defended Avner and told the commander that he was a good soldier. But, if I did that, the commander might focus his wrath on me. He might put me next in line to be dismissed. It was not worth taking the risk.

What a pity that Avner was dismissed.

I feel badly that Avner is leaving, but if he leaves there are fewer people left in our crew and my odds of completing the training improve.

I am glad that it was not me. He did not dismiss me.

Maybe the commander was right and Avner is not a good enough soldier?

I must avoid challenging this commander since he may put me on his list of expendable soldiers.

Similar thoughts passed through all of our minds but no one dared to approach the commander and say anything. The fear of getting on his bad side pushed away any thought of opposition.

Six months later, after we had been an operational crew for some time, all of us would have fought any suggestion of dismissing any one of us. We fought for Dany and for Shay. But, when we were still in the training stage, we felt powerless and impotent. We believed that dismissing Avner was wrong. Inside we knew it was unfair but we were blinded by the pressure and could not see beyond that. We were all too afraid of being dismissed.

All of us felt fear. As the weeks passed, the nagging thought it was wrong to dismiss Avner grew stronger. With time we began to have stronger doubts about our commander's ability to always make correct decisions. We came to understand that we needed to be on guard and protect each other from the whims of this commander.

As we progressed in our time of service in the Unit, the gap between the decisions made by our commander and what we thought were the right decisions, grew larger. There was no possibility to bridge this gap. Avner's dismissal was a milestone for us.

After he left, we all felt a sense of guilt that soured our faith in the command structure. We asked ourselves, *Why didn't we protect Avner and fight for him?* In hindsight, we all know now that we should have done so. Today we all consider Avner an integral member of our crew and invite him to participate in all our gathering and activities.

Eighth Parachute Jump

Thou shalt not be afraid of the terror by night,
nor of the arrow that flieth by day.

(Psalms, Ch. 91, V. 5)

We were told "Night parachuting is a piece of cake."

We were told "You don't see the ground, so there's less to fear."

I was afraid. I prefer to see the ground. During the daytime parachute training, I enjoyed jumping out of planes but six months had passed since our last jump, and now we were going to jump at night. That was scary.

My fear of heights is awakened whenever I am on a high point and can't see a way to climb down safely on foot. It is a sharp sensation of energy flowing from my groin up to my throat. The intense training quieted the anxiety, but with each break it started to build again. During the three weeks before the night parachute training was scheduled, each passing day made my fear more real. I couldn't explain why I was afraid, but my fear escorted me everywhere like an unwelcome companion.

The day of the night parachute jump finally arrived. We attended a briefing in Tel Nof, the Airforce base with the parachute training center, that afternoon. Everyone who was going to jump was given a sack to hold his gear. We were told to tie the sack to our leg with a rope. The idea was that the sack would hit the ground a split second before us so we would know exactly when we would land. We practiced all the maneuvers we would need before, during and after parachuting. Standing in the airplane, walking towards the open door, jumping out of the plane, free falling, opening the parachute and what to do in case the parachute didn't open, gliding down and landing.

What if? If the airplane crashes? If the parachute does not open?

Each of these thoughts forces a chill through my body. My thigh muscle is tight. Breathe. Breathe. Some of the tension is released. Try to think of something else. Start to breathe heavily again. Tighten the rope around my leg and my shoe laces. Concentrating on these activities distracts my attention and lowers the anxiety.

We climbed into the Nord airplane* after sun set. The Nord took off. I tried to quiet my thoughts enough and take a quick military nap. But the noise is deafening. No talking can be heard. Talking or sleeping are the two activities used most often to quiet our anxieties when no action is possible. After a few minutes, we arrived at the Palmachim sand dunes. A loud alarm announces that the green "jump" light is about to go on. We stand up. Wait for the green light to appear above the exit door. We drag the heavy sacks towards the open airplane door. My fear rises and a surge of anxiety forces me forward. The green light is on. Others go before me: "Jump." "Jump." "Jump." "Jump."

I am standing at the door. I'm outside the plane.

The wind hits me. I look straight ahead. Long seconds pass. The sack pulls at my leg. The wind whips every part of my body and the sensation is different for every limb. When my parachute opens, I feel relief for a few seconds. The night is clear and the moonlight illuminates my surroundings. The ground is getting closer, at first slowly, then with increasing speed. I recall the advice of the parachute

---

* Nord airplane is an old piston plane from WW2 that used to be a carrier and became the only plane for training paratroopers.

training instructor: "Tighten your legs." Remember the 'joke' about the female parachutist who forgot to close her legs and landed on a pole. Close my legs. Relax. Tighten. I can't hold my legs together for more than a few seconds. Attempt to estimate when I will touch the ground. Close legs for a few long seconds. Relax. Tighten. Release. Continue to focus on holding my legs together. Relax again. When my sack hits the ground, I am unprepared with my legs loosely apart. I hit the ground with legs apart. My right leg is strained as I fall. I look down and see that my calf bone below the knee is out of whack. I touch gently and feel an unnatural bulge pushing out the skin. Dislocated bone. I lie there not feeling the pain yet. I see Avner walking near me and call him over.

He comes to me and says, "Let's go to the truck. Everybody is already there."

I lie on the sand and ask him to pull on my right calf as hard as he can. He says, "Are you sure?"

I nod, gritting my teeth, and he, with no further hesitation, pulls forcefully on my right leg. The bone makes a crunching sound as it snaps back into place. I feel immediate relief but soon start to feel the pain. I lean on Avner and limp back towards the truck.

We tell our buddies what happened and they are skeptical although my limp is real. I do not remember if I had an X-ray or went straight to my tent when we arrived back at the Unit. I recall that I slept for three days in a row, uninterrupted except for eating and going to the toilet. Those three days were a great rest for me. On the third day, I started to dream. There were plenty of dreams, but I don't remember any of them

On the fourth day I wanted to assess my condition and went out to the asphalt strip to run the two-mile track. While I was running, I felt myself badly out of shape. I did not want to stay at base. My leg worked OK and the pain was bearable. I wanted to be with my crew in the south for more training. The following morning, I caught a ride with a truck driver who was headed that way and joined my crew.

Sinking in the Swamp

Can the rush shoot without mire?
Can the reed-grass grow without water?

(Job, Ch. 8, V. 11)

Sinai.

The landmarks along the way to the Suez Canal are: El Arish, Bardawill, and Refidim. From Refidim one travels on a gravel road to Tasa and the strong holds on the Canal. The command car speeds along. Any deviation from the road will put it at risk of immediately sinking in the marshes. We stop at dusk. I arrive on time and am able to put on my pack and join my friends for the training. We start walking through the marsh.

The north Sinai marshes offered new and unusual smells and sensations. Although we had already covered thousands of miles in Israel, the marsh surprised us. It demands that we move quickly to avoid sinking, yet my speed is curtailed by the heavy weight on my back. My level of fitness limits me to a brisk walk. My left foot sinks in the muck as soon as I transfer my weight to my left leg while attempting to pull my right foot out. The process of pulling one foot out while the other foot sinks in the mud is both frustrating and exhausting. The air stinks like a mixture of rotten eggs and burning garbage. Thick warm waves of the stench hit my nostrils. At times we walk in a boggier portion of the swamp. It takes all our energy just to pull one foot out while the other sinks in. I force my aching leg muscles to extract my feet from the mud. While struggling to go

176

forward, another wave of the rotten egg stench fills my nostrils and lungs. After endless hours, we reach the gravel road and collapse.

The commander orders, "Each three a canteen."

We eagerly drink. Our uniforms are soaked with sweat. My breathing is returning to normal.

"Line up and count off."

We climb to our feet. We move again into the muddy marsh. Pull the right foot up, then the left foot and so on and so on.

My thoughts fade and I am only aware of pulling one foot out of the mud while the other sinks, a process whose seemingly endless repetition makes me feel like Sisyphus pushing the stone uphill. While walking, the only variation from the exhausting routine is the variety of unpleasant smells carried by the wind. My reality is limited to this dismal place where I feel only the sharp pain of my aching muscles. My pulse rate rises and stays high as the struggle through the muck becomes more challenging. I'm drenched in sweat from the effort.

Another break and we gulp down water. Up and walking again. After five or six short breaks, the time is well past midnight. The concept of "Choshech Mizraim"* becomes clear to me as I stare at the back of the man ahead of me whom I can barely see in the darkness. The stink of the swamp and oxygen-starved air make it hard to breathe.

We are told that this break will be a long one—20 minutes. We will take turns standing guard. The two officers leading us move 50 yards away and look through their binoculars toward Egypt. Each of us attempts to find a suitable rest spot for the "huge" break we are given.

I start to nap, my breathing slows and I begin to dream. In this reverie, I see myself in the room of an attractive girl at my kibbutz.

* Choshech Mizraim means a deep night darkness where one barely can see one's own palms. Choshech Mizraim refers to the darkest darkness.

We start caressing each other. Strong erotic sensations course through my body and I am swept up in the world of my dream. My hand moves from her shoulder to her breast. We caress and kiss gently, feeling like we have all the time in the world. The pleasure flows from the center of my body to my toes and fingers. Nothing exists in the world except these gentle, pleasant, erotic sensations. Swinging slowly together in a hammock, gently touching and moving in sync with the slow beating of our hearts. Our breath mingles softly and the moment seems endless.

I am awakened by a warm breeze and abruptly pulled back to the reality of our training existence. I am back on the road side. The few moments in the erotic realm of my dream seemed like hours. Break is over. We are back slogging our way through the muddy swamp. The motion of walking is mechanical and exhausting. I continue to pull my feet free from the shackles of the mud. The leg movements are unvarying. My tiredness is greater and yet, I am in a different realm. The delight of the sweet, erotic dream stays with me through the depths of the north Sinai swamp, and I remember its pleasures during the few hours of the night that are left.

We arrive at an outpost on the Canal and fall asleep, gathering our reserves for another long night of walking through the swamp.

As we get ready for another night of walking, I cling to the sweet residue of the erotic dream from the previous night that carried me through the north Sinai swamp.

Sexy Phone Voice

As far as we were concerned, IDF time lasted for the work week, but beginning Thursday night, returning home from the Suez Canal, was "our time." The longer it took to get home, the shorter the vacation. The pressure to get back to the Unit base and from there head home was enormous, and everyone felt it. No time for mischief and detours. None of us wanted to stop for a break. We were as of one mind, thinking: *Go, go, go....*

The feeling heading back to the Suez Canal on Sunday after a weekend at home and being back on IDF time was one of complete surrender.

On one of our missions, we were assigned to an outpost near the center of the Canal. For us it was a rest break from the exhausting training missions we had just completed in the north. Each man claims his small corner, adapting to the outpost's space limitations. We set the rotation order for guard duty and get organized. Our routine consists of morning gym, cleaning gear, taking our turn at guard duty, playing Backgammon, and eating meals. As the sun sets, the darkness settles heavily around us. Only a few lights blink on the western side of the Canal. At night, darkness obscures every corner of the base, broken only by the headlights of a military vehicle entering the post and parking in the yard.

We all kept "night discipline" while on guard duty. Even Ben Ami waited to light a cigarette until after his turn at guard duty was done and he was within the post bunker. A soldier on guard duty has all the time in the world. The differences were striking between our current, "easy going" situation and the extreme exertions required during our night climbs on Mount Hermon.

During those long, idle nights on guard duty, Doron discovered he could contact home using the military cable lines. We tapped the IDF telephone line and created a circuit to the Baluza switchboard that served the whole Canal area. We asked for a line and the operator connected us to our friends back home. It was a miracle talking with civilians living in another galaxy. Doron had the talent to convince the switchboard operator to provide us with a line for "non-operational" purposes. Within two days, we had figured out her shift hours. Our crew, a band of young soldiers, were aroused by the sound of her sexy voice. We connected two additional phones to the main line and listened in on Doren's conversations with her. We would wait for hours just to listen to her talk for a few minutes.

We were hypnotized by her voice. Each of us fantasized about her at the telecenter in Baluza an hour away. We fantasized about her and compared notes—legs, hands, face, eyes, breast, butt and so on. Her appearance became more and more real to us. With each conversation, we felt that we were getting to know her better. In the middle of the sea of darkness around us, her voice over the phone line lit up our nights and penetrated into our dreams.

Doron, the initiator of the conversations, decided that he had to meet this switchboard operator. Despite the desire not to linger an extra second on our way back home for the weekend, we resolved—Doron, Lony and me—to stop on the way home at the switchboard bunker in Baluza. Doron found out the girl's name the night before we were leaving. We traveled from the canal post to the base in Baluza and began our search for the "voice."

The strength of our fantasies was as powerful as our imaginations. They drew us towards the switchboard like iron filings to a magnet. Doron leads us. Where is she? Other female soldiers direct us to her. We see her, smile imprisoned in an obese body, and...

Fantasy is crushed by reality.

Doron mutters a few polite words to her and we turn away. It's hard to describe the power of the letdown when reality confronts fantasy. We run back to the command car and drive on fast to gain back the precious few minutes of time missed at home by blindly pursuing our fantasy. "You shouldn't try too hard to realize your fantasies," says Lony, concluding this chapter. Speeding home.

Recreation in Jerusalem

As part of our training, we learned to navigate using aerial photos to guide us. One of the first practices was night navigation with aerial photos in Jerusalem. We called it "roof navigation" because one sees rooftops of houses in the aerial photos. Classic navigation is based on reading terrain features such as hills and valleys. Aerial navigation makes use of landmarks of the terrain such as houses, trees, electric lines and roads. It enables soldiers to move with high precision. At the time, aerial navigation did not seem challenging enough for "special navigators" like us—thinking so highly of ourselves. We did not know how important this skill would prove for us less than a year later when we started intensive training in the Golan Heights. At that time, aerial navigation seemed like child's play to us. Going to Jerusalem was a lark for us who were accustomed to covering over 25 miles a night.

During those navigations, we got used to the special smells of Israeli vegetation such as hyssop, sage, wheat, barley, citrus and seasonal fruits. We learned to listen in the night silence for the sounds of crickets, nocturnal birds, the mountain winds blowing, the noise of accelerating car engines and dogs barking. We had been walking all through the night in open, dark fields until the break of dawn. Encountering navigation trails filled with lights, houses, paved roads, cars, people and everything else that exists in the city was surprising. Suddenly, I am walking around a city on a very short, enjoyable trail. We jump from roof to roof in the old city of Jerusalem. Accidentally, we penetrate the intimate spaces of the city and quickly retreat. I am focused on the navigation trail. Do it precisely and quickly. Pay no attention to the fact that I am interrupting the private routine of the city's residents. A dog is barking. I hear music. The musty smells of ancient buildings stone, sewage and cooking swirl in my nostrils. What a new and exciting experience.

Lony, who grew up in Jerusalem, instructed us to meet at the Angel Bakery, which stays open all night, after we finish the navigation trail. In less than three hours, we assembled at the bakery, finding our way by following the smell of fresh baking bread. Each of us got some. Delicious. We shared our experiences about hopping from roof to roof, climbing down into closed yards and passing through narrow allies.

At the end of this trail, a gift. No more walking for the rest of the night. Most of our crew climbed into the truck heading back to the base. Because Lony's parents lived in Jerusalem, he, Doron and I, who shared room number 2, got permission to stay until the next morning.

Among crew members, strong connections are created. Among roommates, intimate connections are built. We would discuss army life, talk about our parents, girls and friends, and share past experiences. With each passing week we got to know each other better: what irritates us and makes us happy. My friends could read my mood without me saying a word.

We visited Lony's home. A large, beautiful house in the Rechavia neighborhood. Lonys' parents welcomed us with warm hugs. Then they left us alone and we sat in the spacious, beautifully decorated living room on three comfortable sofas, and started to talk. Lony brought out three large snifters and poured each of us some cognac.

"At first, let the liquid swirl around slowly in the glass and smell," he said.

We breathed in the rich aroma of the expensive cognac and began giggling like children. We felt good. We talked and occasionally took a small sip from our glass. Time passed. One glass of cognac was enough fuel for an entire night of conversation, stories and laughter.

When dawn broke, Lony suggested that we visit the Sami Bourekas store* on our way to the Unit. Heading towards Lod we did and experienced the wonderful taste. No question. they were the best cheese and spinach burekas on Earth. I ate. The heat, color, sesame seeds and crispiness experienced from the first bite was without comparison. Never to be forgotten. The second bite verified that this slice of heaven on earth was not merely an illusion. We ate slowly and calmly, relishing the savory taste. As we traveled back to the base, we continued to talk while we ate, feeling that life is beautiful. Even Friday check could not erase the taste of a wonderful night in Jerusalem that began on the rooftops and ended in the alleyways of Lod, sampling wonderful bourekas.

* Sami Bourekas is a famous shop of savory pastry shops. Bourekas are pastry made with filo dough and stuffed with cheese, spinach, potatoes or meat.

Great Desert Escape

And David fled and escaped that night.
(Samuel A, Ch. 19, V. 10)

The doctor promised that we would not feel hunger during the three days of exercise in the Sinai Desert. "Just water. Drink a lot of water, and make sure you stay well hydrated," he said. "Fill your cells with water. Pay attention to the color of your urine. Make sure it is as clear as possible."

I knew my body and its capabilities from thousands of miles of marching and navigating. I did not know what would happen when I went without eating for a few days. Each one of us carried a tin box that contained a few candies and a tube of milk concentrate. I was afraid of becoming dehydrated. We discussed the warning signs of dehydration so we would recognize the onset. But the real danger was our ability to deny the reality, even in the face of obvious signs of trouble. We were trained to deny any sign of weakness.

We flew in a Sea Stallion (CH-53) helicopter. We drank water and freshly squeezed orange juice. We kept drinking during the whole flight, which was not long. We landed somewhere in the middle of a large patch of land in the Sinai Desert. The desert sun sent its last rays towards the skies saturating the scene in pink that soon transitioned to blue and then darkness. Flat open space. A distant mountain range to the north dimly visible. The mountain ridge faded away, devoured by the darkness. We drank up until the last second on the helicopter and then descended carrying little weight. An Uzi, cartridges, and plenty of plastic canteens full of water.

We piss clear liquid.

By the time the helicopter disappears eastward, the last soldiers to arrive are ready to start the trail. Adi, Dany and I get organized. Piss again. We look at each other and communicate without saying a word: "Go!"

As usual, we start marching at a rapid clip. I arrange my gear while walking. Move the machine gun strap forward. Adjust the army belt. Routinely feel the pouches on my belt with my palm ensuring they are tightly closed. We stop and everyone looks at his compass to be certain we are heading north. Check with each other—nods of approval from all three—and on we go. Accelerate. Our steps cover more ground. Enter into our rhythm.

We do not know where we landed, but it is clear that moving north will take us within the drifting fan of the larger wadi basin. The mountain range we saw when we landed will escort us a bit west of our trail. After an hour and a half of marching, we stop for a short break. We sit. We are covered with sweat from our fast march. The aridness of the desert dries up the sweat immediately. Per our routine, the three of us drink from the same canteen. Not a word. We get up and resume our march north. Find our rhythm again. No navigation challenges now. I look for the fastest, easiest trail forward. The southern terrain is comforting. I recall the harsh demands of the northern terrain, full of sharp basalt rocks. Here there are just a few stones and soft airy loess soil. The small amount of precipitation does not create much mud to imprison our boots.

We continue to march briskly. We stop every hour and a half for a water break—three to a canteen—and immediately move on.

Despite the effort to maintain our rapid pace, tiredness begins to slow us a bit. The goat trails are comfortable and clear on the flat areas, seeded sparsely with flint stones. Still moving fast. But our pace is slower than when we began the trail eight hours ago. Everyone is focused on his own thoughts.

I lead and a song plays in my head: *Blue shirt, blue shirt, outshine any ornament. Blue shirt, blue shirt.*

And this song is replaced by:

What is at the end of the ascent?
Descent, descent is coming.
Come, come, come.
What is at the end of the descent?
The ascent, the ascent is coming.
Come, come, come.
What is at the end of the ascent?…

Go down into a shallow wadi bed. The attention required while descending pushes the invigorating Machanot Olim youth movement songs from my thoughts.

Looking ahead, I search for a familiar landscape mark. Find nothing. Know clearly only the direction north.

The people of Israel might have passed through here on their way out of Egypt walking through the desert to the promised land. Probably not too far away, Moses struck the rock for water and was punished and not allowed to accompany his people there. Oh, the twelve spies might have left the camp around here and journeyed east to inspect the promised land.

The desert aridness dries the sweat on my body and I barely feel it. Only the parts of the army belt pressed against my body and not exposed to the breeze remain moist. I feel the cool wetness only after stopping for a few minutes. Move on. We try to drink as little as possible so we will have enough water over the next days.

Walking in the wadi's bed, I smell fresh dung or hyssop under my boot. I stumble near the root of a juniper brush. To avoid falling, I run forward three steps and resume my walking pace.

Marches of hundreds of miles alone—through rain, fog and clear nights—have given us a sense of confidence in the dark. The night becomes a familiar companion. After so many nights of navigation training I felt very safe in the darkness. I felt that at night I can protect myself, we can protect ourselves, no matter what dangers confront us. With each mile of walking our confidence grows.

I recall the visit of Mordechai Piron, the IDF's chief Rabbi. We gathered in the Unit club where pictures of soldiers killed in operations were hanging on the wall facing us. With much emotion he said, "You. You are executing the most important Mitzva* in the Torah. You risk your lives for the protection of the people of Israel. By your deeds and devotion, you are protecting Israel."

His statement sounded strange to my ears. I felt proud to be part of activities that were contributing so much to the security of Israel. But my actions had nothing to do with Mitzva.

He continued, "You are heroes to the people of Israel. You are the ones continuing the glory road of Bar-Kokhba."**

I felt alienated. The legacy of Bar-Kokhba does not speak to me. I thought *What did Bar-Kokhba do for Israel? Brought a national disaster? To what I can connect?*

---

* Mitzva is an instruction in the Jewish faith. There are several Mitzvas held of high regard such as giving to the needy and protecting lives. Related to this Mitzva is a proverb—"He who saves one life, it is as if he saved the whole world."

** A Jewish military leader who led a revolt against the Roman Empire in 132 BCE. Its failure was a disaster for the Jewish people. A substantial number of Jews living in Israel were murdered or enslaved by the Romans.

That question stayed with me. I remembered it dozens of years later when I began studying Judaism in Kolot. Then I came to understand that my Jewish hero is a different leader, Yochanan Ben-Zackai. He famously stood in front of the Temple in Jerusalem and scolded it! And also warned that adhering to religion's external trappings and hardline interpretation of Jewish law would lead to idolatry and destruction. After the lengthy siege of Jerusalem by the Romans, Yochanan Ben-Zackai humbly asked the Roman general that just the Yavne village to the west and its "wise people" be saved. The general granted his small request which enabled Jews and their culture to survive in Israel and, later on, in the Diaspora and to live, learn and prosper.

We continue to move northward during the final portion of night. While walking, I feel heat and cold at the same time in different parts of my body. My chest and armpits are warm and my fingers are almost frozen. As dawn approaches, the wind gets colder. The joints of my fingers are frozen. Even a short break causes my legs and back to feel cold. I start to shiver. We get up and continue our march.

In a little while, we will start getting organized to pass the day hiding from the Bedouin trackers searching for us. We will not reach the mountains tonight. The first rays of light are spreading to the east. Streaks of red and pink color the sky. A short piss break before we find a hideaway. The western sky is still dark transforming into dawn. We have to find our hideaway in the next few minutes. We spread out and search for the right spot, moving northward. In the distance, we see a large dense, circle of juniper bushes. We locate the entrance to the hideaway 50 yards before the bushes, through a hard terrain path. We walk in a large semi circle southeast in the soft wadi bed and climb onto the hard terrain leading us to other large bushes. I tear a branch close to the ground. Adi and Dany go ahead while I walk behind using the branch to erase our footprints. We move into the center of the juniper bushes and settle in.

We arrange ourselves around our gear. I attempt to find the best position to protect my body from the cold. I wish I had a blanket to cover my sweaty body and protect me from the freezing wind at this early dawn hour. My cold, wet shirt sticks to my body and sends waves of shivers through me. I get goose bumps on my legs and arms. It is so cold that I can see my breath as I exhale. I want to walk around to warm up and find a good place to sleep so I can rest my tired muscles. I place the rigid radio under my head as a pillow and curl up in a ball like a fetus. We all curl up to preserve our body heat. The inner parts, belly and chest, are warming up, but our backs are exposed to the cold wind. The doctor told us that we have fewer "cold sensors" in our back. I lie, back to the wind. Adi is near me, his back facing me, and Dany's back faces Adi. My sweaty shirt continues to dry while chilling me. Each gap exposes a few inches to the cold wind. I think *Why didn't I bring a piece of nylon to protect me?*

Fall asleep. Wake and fall sleep again. Penetrating rays of light signal that the sun is starting its daily trek across the skies. Continue to doze and wake, doze and wake in an endless cycle. The light is getting stronger. The wind diminishes. More sun rays penetrate the bush, gently warming us. The periods of dozing increase from a handful of minutes to 10 or 15 minutes. The desert warms up slowly. As the temperature rises, my body relaxes. My breathing slows down and my skin warms up. I take a sip from the canteen. Everyone relaxes as the sun warms us. We manage to get a whole hour of uninterrupted sleep. Cover my eyes with my hat and spread out on the sand. As the desert heats up, the shade of the bush is pleasant. We can enjoy ourselves resting within the circle of bushes until evening. We are replenishing our energy after the intense march. Near noon we drink again. Since there is no food, we do not think about it. For now.

We get ready for another night of marching. Check our gear. All is there. Last light. Set our marching quickly north. Breaks are short. My body hurts. My legs and other muscles are tight. I do not study the map, just push northward. We stop each hour for a quick drink from the plastic canteen. During the short break, my muscles don't cool down, so the continued walking doesn't entail too many extra pains. Despite the chilly night air, we sweat. We keep up our fast pace. Enter the zone again, where our brains are disconnected from our bodies. Recall that we haven't eaten. Remember my dad.

On April 11, 1945, at 10 in the morning, a tank broke through the fence near the camp gate to Buchenwald. Then another tank and another. A large black soldier stood up in the tank turret and shouted, "YOU ARE FREE!"

"Like a voice from heaven," Aba recalled 60 years later, when we visited Buchenwald with him. Block 66. The "children's block."

After the tanks came food trucks. A spring day-dream. The skeleton that weighed 32 kilograms took in the sight he had waited for endless, dark years. He did not move. He could not move. He was completely exhausted.

Other prisoners fell upon the food that was offered and, despite the warnings of the Americans, gorged themselves. To death. Aba was too exhausted to move.

An American soldier, well fed and warmly dressed, saw him. He carried him to the hospital building, which had been the headquarters of the SS guards, the essence of evil. Dad burned with fever. The doctor, an American Jew, offered him a new, life-saving medicine. The antibiotic came in large white capsules. "Take three pills a day, morning, noon and evening. Within a week you will start feeling better."

"OK."

After a week the doctor came to examine him again. Following the examination, the doctor declared, "You are healthy. You see what a wonderful job these pills perform?"

Dad smiled, placed his thin arm under the pillow, and said, "Thank you doctor. These are the pills that saved my life?" And handed the doctor a handful of white pills.

I laugh out loud. My crew members look at me and I smile back, nodding my head—never mind.

We push northward. Walk with more energy.

A song about the 1948 Independence War enters my head:

Bom, Bububom, Bububom.
The whole world is darkening…
The city is dying under siege.
No bread, no water, nothing…
Thirty at Nebi Samuel.
Seventeen on the Kastel.
The Nebi Daniel convoy…
But we will overcome…
Open the shutters…
Since in front of your house they pass.
Those who maybe, maybe, maybe will come back.
Bom, Bububom, Bububom.
The whole world is darkening…"

We walk. There is nothing romantic about the frustrating existence of the walker.

We called ourselves "Golda's suckers."* Just say when…and we jump! We all took tremendous pride in being allowed to serve thus. To be part of the military elite. The leaders. To be "Golda's suckers."

---

* Golda Meir was the prime minister at that time, and the phrase meant we were willing and eager to serve the State of Israel.

At the end of this night, we hide again within another thick clump of bushes. The wadi is wide open. We have plenty of water but we need to set some aside for the next night. Only one more night. Then, we will call the helicopter to pick us up and carry us back to the base. The flies swarm around us. *How do they find us so easily in this vast dry expanse of desert?* I cover my face with my hat. Raise my collar. Only small cracks are left for the flies to penetrate. Doze. The heat rises as the sun ascends. Drink water. Much water left. We were parsimonious. Only one more night. Cover myself and fall asleep.

Day-dreaming. *What pushes me so powerfully? Dad?*

In his body he held in the anger, frustration, pain, sadness, oppression, howls of anguish. And channeled the screams that were not cried out into each cell in his body and passed them on to me as a trembling rage. Wounded, inflamed, pustular, uncontrolled. Propelling me through the endless, uncompromising, demanding endurance marches. The pained, trembling shouts. Only the exhausting marches can calm the uncompromising demands. Are there other voices?

Isaiah: *I gave my back to the smiters.... I hid not my face from shame and spitting.*

I can't connect with Isaiah. I revolt at the concept of "Turning the other cheek," part of the teachings of Jesus and Isaiah. I am not going to allow my back to be whipped or turn the other cheek. The ability to absorb the counsel of Jesus and Isaiah, to placate the anger and aggressiveness do not belong to me. I am very far from that place.

I want to be strong. Invincible. Cruel when necessary. I am happy to be a unique instrument so that Israel will stay well ahead of its enemies in this dangerous part of the world so we will never experience the bitter cold and hostile alienation that Jews endured in Europe. That place that shaped Mom and Dad. And myself.

Wake and doze. Wake and doze.

A long night lies ahead of us, but at the end of it the helicopter will land and fly us back to our base, where we will have a briefing, Friday conversation with the commander and crew members and home. Recalling the smells of home and the approaching weekend remind me of their existence.

Getting ready for the night march. We carry over half of the water in many canteens on our army belts. The weather is comfortable. My stomach reminds me it's empty, but I am not bothered by it. As soon as we get on the helicopter, we will be given food. My energy is channeled into walking. To get quickly to the pick-up point. North. North.

Marching briskly again. The walk is monotonous and we move quickly on the plateau. The night is clear. Several acacias and junipers are visible in the wide wadi. We all make efforts to walk as quickly as we can and be the first to arrive at the pickup point. Adi and Dany pull ahead and I join them. After two hours we stop for a hasty drink. Dany repacks the empty canteen in his army belt.

We cross a stretch where water remains standing for a few days and hear "krachh, krachh" for a few minutes as we tramp through the "chocolate pieces" of dry loess land. As we move past the dry puddle, I recall a series of beautiful images of the 'chocolate pieces' in dried puddles in the Negev. In some of the pictures, fresh grass grew between the cracks. We climb to a plateau topped with small flint gravel. Our boots make a sucking sound as they sink into the sand of the loess and hit the flint stones. I look up and see the North Star in its place in the sky, marking the direction. The dry air is redolent with the scent of a shrub that one of us stepped on with his boot.

Water break. All of my leg muscles are weak with exhaustion and the soles of my feet are burning. I could easily lie down and rest, but my friends "pull ahead" and I have to stay with them. No time for rest. We want to be the first to arrive at the landing area. All the other squads are also marching quickly. Although we did not talk, everyone knows that this is a competition among the squads. I do not pay attention to the increasing pains in my legs and feet. Very focused on getting ahead of the other squads and being first to arrive at the landing.

Five hours remain. Three. One. We stop for a final water break. The last sips are taken while we walk. Place the empty canteen in my army belt while walking. Close the 30-yard gap, I start running. Getting closer to the evacuation area. Listen to the radio. Climb up on the northern wadi bank and walk for a few 100 yards north. The area is flat, sown with small flint stones. We place the four small torches in a trapezoid shape against the wind. "Ready to evacuate" we report to the approaching helicopter. Are we first?

Morning fog starts to roll across the land. We hear the other squads declaring "ready to evacuate."

After us, I say to myself.

The helicopters pick up the other squads before us. The last helicopter is about to pick us up and then after us, the commander's squad. The Sea Stallion is descending as it approaches the landing site. We get ready to the lowered helicopter's back door and to run in. Turn our back to the helicopter so we don't "eat" dust from the spinning rotor blades. The torches are extinguished. The helicopter touches down … and then begins to rise again…

We hear the commander insists that his squad be picked up first. What is happening?

The pilot says that he can only pick up one more squad because the fog is getting dense. I am furious. Adi lifts his Kalashnikov rifle and shoots a bunch of tracer bullets in the direction of the disappearing helicopter. I look at him in amazement. He says "I didn't aim at them." I know.

Adi and I are cursing, "Kus emek ya Hara!" Furious at the commander who insisted that he be picked up ahead of us.

Dany calms us. Maintains a philosophical detachment. They contact us and say that we will be picked up by a "terrain helicopter." The command car will arrive within three hours—with food. Adi and I are still raging. We didn't take into account that the commander has to prepare for the next week of training.

The wind picks up. We walk south and hide in a small ravine on the wide wadi huddling together to stay warmer. Doze off. We each take turns standing guard. The wind is getting stronger. I start to shiver. We all shiver.

First light.

The command car driver reports that he is getting closer but doesn't know exactly where we are. We shoot off fireworks. He says that he saw them and will arrive at our location shortly. We see the command car and direct him to us. When it arrives, we load up our equipment and climb in.

We cover ourselves with thick coats. The command car pitches and lurches going through the deep tracks of the wadi. Finally we are in sight of the asphalt road. There is plenty of water in the car and we still have almost half of our canteen water. Dany opens a field ration and starts eating. Adi and I decide not to eat until we arrive back at the Unit. Since we managed to get this far, stretch our ability to go without eating for a bit longer. The experienced driver navigates quickly and easily through the challenging wadi terrain. Every once in a while, we are surprised by a hard lurch when a tire hits a rise or a groove in the track.

Drink again. We reach the asphalt and stop the car to piss. Our urine is a healthy light yellow. Drink and climb back into the command car. The driver disengages the four-by-four gear and we make it up onto the asphalt road.

We "compensate" ourselves with a trip to the beautiful Kadesh Barne Oasis. Long nap on the way back to the Unit. Arrive at the base in the afternoon. I feel good that we refrained from eating field rations. Abundant fresh food awaits us.

Stay Awake Competition

As in many situations, deviation from the established path—a word, an act of mischief—can lead to the imaginary tunnels of *Alice in Wonderland*.

"You are sleepy."

"Look who is talking."

This exchange led to a competition: "Who can go without sleep the longest and still stay alert."

The competition began on Wednesday, after one of our night training missions on the Golan Heights, just before we spread out to sleep on our IDF mattresses.

We, the other crew members, cheer on the competitors and place bets. Some bet on Adi and some on Ovad. They will compete to stay awake while the rest of us are stretching out horizontally on our sleeping bags. We quickly drop into a deep sleep, resting for a few hours before our morning tasks commence. Ovad and Adi sit up, drinking warm tea, talking and gently teasing each other. Both are determined to stay awake.

Adi was dismissed from a pilot course and joined us. He is the only member of our crew who was never off by even one yard during our dozens of navigation exercises. Ovad, a talented athlete, is always one of the three runners to arrive first at the finish line of the one-mile Sunday race.

We are deeply asleep. As one of us wakes up to go piss, he looks at the two competitors and mutters, "The two of you are already asleep."

Both reply, "Not me. He is asleep."

We continue to slumber in our warm sleeping bags.

We complete the Thursday morning tasks while Adi and Ovad continue their competition. With each passing hour, the look of exhaustion on their faces deepens. The "friendly teasing" between them continues as the competition goes on.

"Adi, you closed your eyes."

"Ovad, open your eyes."

Each comment starts a debate between the two as to who fell asleep.

We get ready for the Thursday night march. Both men are very tired. The walk is monotonous and the work is familiar. The night reaches its end. Light snow is falling. We climb on the bus that takes us to the base for Friday's briefing and then home for the weekend. The sleepless competition continues. Adi tells Ovad jokes, and then Ovad tells jokes to Adi so they can stay alert. The bus moves slowly on the paved road, which is covered with a thin layer of snow. The snowflakes strike and then slide down the front windshield of the bus. I place my face close to the glass and imagine I am in *2001: A Space Odyssey*, a traveler between galaxies.

A comment between the contestants, "Why be tired on Friday and Saturday?" leads to "Well, the struggle does seem pointless." Which then leads to, "Let's both fall sleep."

And so it was.

Back to the Dreaded Hermon

From Aroer, which is on the edge of the valley
of Arnon, up to mount Sion—named Hermon.

(Deuteronomy, Ch. 4, V. 4;8)

On April 10, 2003, at 6:40 a.m. we gathered, Amnon, Yadi, Ovad, Hagai, me and family members, at the lower station of the Hermon funicular to set out on a bicycle ride from Hermon to the Mediterranean in memory of Shai and Dany. At 7 a.m. we climbed on our bicycles and headed down.

Riding down the Hermon's narrow paved road, memories from three decades ago surfaced. As we navigated the curves, the chilly wind mixed with the warming rays of the sun. I remember, smiling: *How many times had we walked up and down this mountain on the sharp rocks far from the paved road? How many times had we traveled aboard a D500 truck on this road, from the center of Israel to the Hermon, while resting on our IDF mattresses, attempting to take a nap?*

Many times we reached Nafach* in the Golan Heights and took up position in a run-down, deserted house, protected from the wind. The cook prepared an early dinner for us consisting of French Fries, salad and roast chicken. We ate the meal while our thoughts drifted towards the anticipated ascent. We added chocolate bars to our army belts. Most of my crew members liked milk chocolate while I and a few others preferred bitter dark chocolate. We shared three

* The remains of a village destroyed during the 1967 War's aftermath. One four-room house had its roof still intact, and we used to eat and sleep there during our weekly training in the Golan.

chocolate bars for the night. In addition, I had brought some dried figs from home and shared them with my friends. I kept some for myself to eat later on the harsh cold trail.

We were on our way to the Hermon Mountains. An hour before sunset, we climbed aboard a D500 truck and traveled along the narrow roads of the Golan Heights. While dozing off, I rehearsed the steep, ascending mule trail in my mind, climbing the Hermon with ease. I mentally reviewed the list of gear to be certain we had everything we would need. We inspect our gear multiple times before loading our equipment on the truck to ensure the likelihood of missing anything is slim. However, checking that the water is fresh and that we have chocolate bars is not part of the inspection. These are details that affect the arduous experience of climbing and can make the experience a little better. Each crew member is enclosed in his own private world. Nobody wants to talk. Especially not now. The truck stops.

"Get down. Gear on." The harsh familiar order from our commander.

We climb off of the truck gloomily as is part of our routine and stack the equipment on the road side. The chilly Hermon wind reminds us of where we are. The wind carries a sharp smell of Druze fire. The sounds of human voices and dogs barking can be heard from the nearby village. I smell the earth, the local vegetation and feel the cold. I sense all these in small measures, knowing that the force of the wind and bitterness of the cold can alter dramatically in a split second. I carry my army belt, weapon and a carrier. We all have one, loaded with dozens of kilogram weights, carefully balanced for the long march. Each of us spent hours preparing our carrier for the long hike. The two carrying straps, pressing against my shoulders, are covered with a wide spongy material to distribute the weight as widely as possible and delay the sensation of the straps cutting into my shoulders. Despite their additional weight, the army belt and the rifle did not count as part of the burden.

During our preparations at the base, we all 'brainstorm' how to reduce the weight of our backpacks. Hagai is the most diligent. If he recognizes a small screw he thinks is unnecessary, he takes it out. To our question, "what is the point of removing one gram?" He replies: "Every gram I leave behind is less weight and leaves me with more endurance for this endless march."

The temperature forecast and weather conditions greatly influence our clothing. We already know that if we are dressed too warmly, after a few minutes of walking, we will feel our body overheating and will need to strip off a layer of clothing that becomes an extra burden to be carried along the entire trail. We also risk becoming dehydrated. On the other hand, if we don't have enough layers of clothing, it will take us forever to get warm and we will shiver during every break and not be able to rest. Thus, proper clothing is of utmost importance.

During winter time, I usually wear long cotton underwear under my regular army pants. During the summer, I wear air force trousers that are lighter and airier. Dressing the upper body is more complex. A T-shirt and an army shirt are the basics. In order to keep our body temperatures in a comfortable range, one needs to wear a sweatshirt or a coat with a zipper. The zipper allows us to start off with a closed coat to keep warm and after a few minutes of walking as we warm up, unzip the coat and allow the heat to go out. We did not have sweat permeable clothing at that time. At times, after opening the zipper of the coat completely, I would fold the edges of the jacket back and stick them under the carrying straps so my chest and belly were exposed to the cooling air. If the rest break was long enough, I would gather the pieces of the jacket around me, cover the front of my body and warm up. At higher altitudes, the cold wind blows so harshly that additional layers of protection are needed. The sweat that accumulates on our skin during the march requires an extra layer of protection from the cold.

Similar to how I wear my clothing, I dress in mental layers. I envision myself covered with an insulating layer like a diver putting on a wetsuit before he plunges into the water. I wear tight socks that rise above my ankle. My foot and ankle are thus protected. Pull on snug pants that stretch across my calf, thigh and pelvis. Thus my leg muscles, knee and hip are shielded. I zip up my "wetsuit" from my belly to my neck protecting my back and shoulders that are straining under the mighty weight.

Now, I feel like all of me is well-covered with protective layers. Insulated from the cold, wind, rain and other forces of nature. I feel like a knight from the Middle Ages. Shielded by my armor. Protected, I go inside. Continue to look inward and enter my own private bubble in preparation for the long strenuous night.

Before we begin walking, we take advantage of the opportunity to drink some warm tea and empty our bladders. We encourage each other: "Within fifteen hours, our lives will improve considerably." This statement causes us to smile. Then, we are back on track focusing on the task ahead. I'm already carrying my army belt and weapon and I look with trepidation at my pack. Positioning my back to it, I squat down. I pull the carrying straps over my shoulders and wait quietly for the commander's signal to begin the march. Notice the smells of cow and goat dung mingling with the distinct scents of the Hermon shrubs.

The sharp smell of a fire from the nearby Druze village assaults my nostrils. The smells and voices from the village raise thoughts of being taken prisoner by the Syrians. When we talked about Syria and the capable Syrian army, being captured was discussed. What is it like to be taken prisoner by the enemy? I read about American soldiers who survived being prisoners in Korea. Soldiers who the Koreans tried to brainwash.

Being taken captive by the Syrians felt like an obscure, ever-present threat. I think of Uri Ilan, a paratrooper who was taken prisoner and committed suicide in Syrian captivity, leaving behind a note: "I did not betray my country." I was afraid of being taken prisoner. Remember David Alon's brother who returned from captivity in Jordan. When he walked up the kibbutz pathway, I could not bring myself to look at him. There was something so strange and different about his walk, appearance and behavior. It was clear that his period of captivity had scarred his soul and mind. I could sense his difference—we all did—although nobody on the kibbutz talked about it.

We heard stories about the cruelty of the Syrians. About beating the soles of prisoner's feet and tearing the flesh. Whippings that left backs scarred for life. Pulling out fingernails, extinguishing burning cigarettes on the nipples and electric shocks to every sensitive part of the body. Drugs that cause nausea, vomiting and blurring vision and distort one's sense of reality. I thought that I could endure beatings and starvation but was unsure how I would deal with the psychological pressures. I was most afraid of drugs. *What will happen to me afterward?* Do not want to think about it. We talked among ourselves that imprisonment in Egypt would be preferable to Syria. Yet, for crews operating in the North, Syrian captivity was a real possibility. Most discussions about it were brief and tended towards dark humor.

We envisioned situations where one of us found himself alone in enemy territory. As a crew, we felt stronger than any force we might encounter. For a moment I imagined myself stranded in Syrian territory after losing my crew. The thought was so frightening I almost jumped out of my skin. Fear of being taken captive motivated me to be well organized and to work seamlessly with my fellow crew members. I knew that as long as we performed our tasks as we were trained to, day after day, the chance of being taken captive were diminished. I knew that any enemy who popped up would be immediately shot down. I knew my Uzi intimately. After hundreds of shooting exercises, under a variety of conditions, I could aim the gun, shoot and hit my target. On the physically demanding Hermon Mountain, surrounded by my well-trained crew, we are powerful and protected. In a worst-case scenario, the IDF will take care of us. Dagan, the helicopter pilot flying the Sea Stallion, will come and rescue us.

Dreams of captivity used to come to me in my sleep, but were usually forgotten when I woke up. My fear of being captured surfaced sharply during brief moments before commencing operation on the Hermon Mountain. Most of the time, the busy activity before operations pushed the fear away.

The rays of sunlight fade away in the west. Darkness thickens with each passing moment. There is no rain but we are walking inside a cloud. At this time, the asphalt road is warm compared to the Hermon rocks. A bit later, everything will be cold. Bitter cold. The light gusts of wind remind us that within a few hours the winds will pick up and the air will grow colder. We are on the Hermon.

The commander clicks his tongue.

I lean forward. Engage the muscles of my legs and arms. Inhale and get to my feet while maintaining my balance. I am on my feet supporting myself as my arms rest on my knees. I exhale and start moving to the trail's starting point across the paved road.

The commander, who carries no burdensome load except his weapon and army belt, is jaunty as he leads us. We fall into the prearranged order behind him. After a few dozen yards, we stop and count off, stating our "iron numbers,"* ensuring all are present. The column ascends slowly. My body temperature rises. I'm sweating. Open the zipper of my outerwear. Cool off. My body strives to maintain a comfortable temperature. During the first 500 yards, I struggle to breathe and am panting heavily. My body transforms from light, nimble athlete to heavy, burdened mule. It's very difficult breathing. Step by step, my body finds its balance. My breathing and heart rate enter a "working" rhythm.

What thoughts fill the head of a soldier carrying dozens of kilograms of weight on his back? Not many. I focus on the sensations of my body and carefully putting down my left foot, then right, left, right. Again and again. During the long hours of the march, correct placement of the feet is critical. Any deviation can cause an ankle sprain. At times, my ankle starts to twists, my back leg shoots forward, and I run a short sprint to regain my balance. I rarely fall. When I do fall, the weight of my pack crushes my body. Falls can lead to small or serious injuries that threaten my ability to continue the march. Falls can easily lead to sprained ankles. All of us at some time experienced such sprains and our ankles were easily reinjured.

To help avoid sprains, before each walk, I wrapped my ankles with elastic bandages for additional protection. I covered the bandage with tight socks and secured my shoe laces tightly. My calf and foot became almost one piece, similar to a ski boot, with minimal flexibility for walking.

* Before a crew starts night training, each member is assigned a number. The last person in line has number 1, and the count goes up to the soldier right behind the commander. At each stop, everyone says his iron number out loud, so it is clear that no one is missing.

As the mule trail becomes steeper, I think about moving my entire leg. Enlisting help from my right calf and thigh muscles. Then left. And so on. I push upward one "stair rung" at a time. Each ascending step is accompanied by a quiet exhalation of breath that becomes more and more pronounced as the ascent gets steeper and longer.

We walk inside the fog and can barely see. The sparse oxygen at the elevated heights contributes to our dizziness. I walk with the goal of sticking to the soldier ahead of me and to the one before him and so on, up to the commander in front. The commander leads his "mule train" uphill.

I barely sense a thing. My awareness moves between my feet and muscles. Each time another muscle group contracts. Then breathing, sometimes listening to my heartbeat. Rarely do other thoughts surface.

The machine gun hangs from my right shoulder, shaking with my arm laid across it. I can't even think what would happen if I needed to use it. My training gives me confidence that if necessary, I will be able to aim quickly and hit my target. Every iota of energy I have is being enlisted for the demanding task of ascending. All that is important is to keep moving up at a steady pace. Place my foot and take the next step. Place the next foot and enlist my leg's muscles. Close eyes. Make the step. Open eyes. Closing my eyes transfers the wasted energy of visualizing to the efforts of my thigh muscles. Like a rest within the walking effort. Just when it seems that our walk will continue forever, we stop for a break.

Immediately, I look for a rock to sit on. Not too many options. Even sitting with my back towards the mountain top, facing the trail, provides minimal room to rest.

"Two to a canteen."

I remove one of mine and hand it to Adi who sits next to me. He drinks and hands it back. Ovad walks from one man to another offering tea. Where does he find the energy to move? Adi and I empty the canteen and I place it in my army belt. One kilo less to carry, I think. My breathing is quieter and slower. My heart rate slows a bit as well. A gap between the clouds allows a few stars to be seen before being quickly covered by the next cloud.

I remember how during one of the rest breaks during Sunday gym I felt my pulse and discovered my heart occasionally skipped a beat. I tell myself that as soon as I start exerting myself, my heart beat will stabilize. A friend from my swimming training told me once that his heart beat is irregular until he starts to strain his body.

A squeak.

I move quickly from sitting to standing while maintaining my balance. It seems that the liter of water I just consumed has no effect. I am rooted to the earth. Start crawling uphill. The pace always seems too fast for me to keep up and forces me to exert extra effort. Start breathing heavily again as we climb. The trail is muddy and I attempt to walk on top of the light grayish rocks. Given the choice between trudging through the mud or risk sliding on the rocks, I prefer the rocks and prepare myself for the anticipated meeting with Mother Earth. My Uzi continues to bang against my side. I try to keep it away from my body. Although I don't succeed, it diverts my attention from a few painful steps.

The trail is very steep now. I can hear everyone's strained breathing. Step, then another step. That is my existence. Keep close to the back of the man ahead of me. The stinky smell assails my nostrils and fades away in the Hermon wind. In time we learned to recognize the smells of our farts along the trail. Distinguish between the smells of pears and boiled chicken with potatoes puree. We recognized the smells of the last portion of our digestive tract—apples, schnitzel, jam and, of course, Luf.* The best, and probably only positive aspect of the Hermon wind was its ability to quickly carry away these smells.

As the night hours pass, our weariness grows. Exhaustion affects muscles groups inconsistently and in no particular order but with inevitability. I can never quench my thirst. The empty canteens do not reduce the burdensome weight I carry by even a gram. The weight feels overwhelming.

After endless hours we reach the top. We collapse on the ground for a short, cold, rest break. I don't berate myself for not packing an extra layer of clothing for protection against the cold. The exhaustion of the climb removes any desire I might have had to add even one extra gram of carried weight. The wind picks up and chills me with increasing strength. Now, during the middle portion of the night, the cold becomes an integral part of us. I don't bother worrying about the difficulties of the descent.

I leave my carrier and go to piss a few yards away. My sweaty back that was protected by my carrier is exposed to the freezing Hermon wind. Return to my carrier to doze for a few precious moments until we start walking again. Eat a dry fig and offer one to Adi. The sweetness is heavenly.

Walking again. Getting ready for performing our tasks. The effort is strenuous. I am breathing heavily. Look at Yadi and make him laugh by exaggerating my effort to breathe. We both smile as we recall the code of "making noise of efforts."

When we worked, Yadi and I, it took the effort of all the muscles in our bodies, as if we were trying to uproot a huge eucalyptus tree. Not like Jacob in the Bible, who easily managed to roll the stone from the mouth of the well, we exerted all of our strength. When we

* Poor-quality, ground meat used in IDF rations. You can imagine how much we enjoyed the smell and taste.

were unable to make headway, we had to show that we were trying our hardest by making "sounds of exerting effort."

I wanted to scream out the strenuous exertions with all my might but we were required to maintain "night silence discipline." We would laugh at ourselves. But every time we arrived at a dead-end, we participated in the game and "made sounds of exerting ourselves."

We finished the task. Now we start the descent. After a few minutes of walking, I am getting warm again. The clouds moved away for a few minutes, allowing a beautiful view of the "Galilee Finger," the northernmost portion of Israel, dotted with villages lights. Stunningly beautiful.

I slip on a rock. The serene beauty is extinguished. My blunder brings me back to the demands of our current reality like the shock of a mule's kick. I regain my balance and refocus on where to step. Pay attention to the placement of each foot. Close the gap to the back ahead of me. Continue marching.

After many long hours of walking, the effects of exhaustion and the thin air make my thoughts cloudy. I enter a new world. I am totally focused on the effort of marching but at the same time existing in a very quiet place. Distanced from the grueling walking effort. Time is misleading. A minute takes an hour and an hour a minute. Time and space are fluid. I am like a machine. A machine programed to get results. The whole focus of my existence is the task at hand. Totally focused. No deviation.

Pains and feelings are uprooted. Sealed away. Dispassionately dissected. Not acknowledged to exist. Not within this time and space. I am placed here to perform this mission. A machine. *How does this machine transform into a human? How does the human become the machine?*

It's almost dawn when we reach the truck. Eat something and get on. Traveling southward. In the truck we cling to each other saying, "Nablus, Nablus."*

We huddle close, chest to back to help us defrost from the Hermon's cold.

As we doze off in the warm truck, on the comfortable mattresses, I think about the cold of the Hermon and the cold in Poland. I do not know the cold winds of Poland. The thick coat wrapped around me was beyond the dreams of the imprisoned Jews. Warm coats were for Germans only. Dad, with no coat, survived Krakow in 1939. Birkenau in 1943. Buna Manovitz in 1944. Buchenwald in 1945. A thin, striped prisoner uniform was all he had to protect him from the cold rain and snow. The chilling wind blowing right through him. Mental pictures of his Death March fill my head.

The tiredness wins and I doze off, traveling south, thankful.

* Our crew decided to use it as expression meaning "to huddle tight and close to one another." (It is also the name of a large city on the northern part of the West Bank.)

At Doron's Home

Every man under his vine and under his fig-tree.
(Kings, Ch. 5, V. 5)

We came down from the Hermon Mountains on Thursday night. Doron invited Lony and me, his room 2 bunkmates, to his home and the commander allowed us three hours away from the base. What a treat. All three of us were looking forward to enjoying the tastes of a home-cooked meal. After the visit we would drive back to the base in Doron's car. As we approached the city of Tiberias, we anticipated exchanging the smell of truck diesel for the smell of Doron's mother's cooking with pleasure. Through the vines around the house, we could see the Sea of Galilee.

Doron's mother welcomed us warmly into her home. She served omelets, vegetable salad and bread to us three "starving" soldiers. We talked as we ate and drank. The smells were pleasant and "homey," and his mother presence was gentle and warm. Her home had a comfortable, harmonious atmosphere. We soon became jolly and light-hearted, enjoying our brief sense of freedom.

It was quite a night. I lost track of time. Delicious fruits kept arriving at the table. This peace and tranquility of Doron's home seemed to last forever.

Finally, Lony says quietly, "Time to move."

We check all of our equipment, making sure nothing is lost or forgotten during our brief "party," and jump in Doron's car.

The hour is late. Warm air flows through the heating system and the radio is playing Israeli songs. I sit near Doron who drives while Lony dozes off in the back seat. We talk occasionally. Name a junction or a village as we pass by. Dead of night. The car headlights illuminate the paved road. Getting close to Hefer Valley. Suddenly we lose sight of the road in a white cloud of dense fog. We both are jolted. In another second we emerge from the fog. Doron slows down. We get close to another cloud of fog and Doron drives even slower. A sea like cotton wool engulfs us. Can't make out a thing. Crawling forward. Barely able to see, Doron clings to the right shoulder of the road. We keep entering and exiting dense banks of fog. Our hearts stop for two long seconds as the road disappears and we heave a deep breath of relief as visibility returns. We alternate between clear dark road and blinding fog. A dreamlike reality. Concentrate on the warmth of the car and the music from the radio. Doron slows down again. We oscillate in waves of clear paved road and blindness. Get accustomed to the rhythm. From Beit Lid to Raanana, the fog gradually thins and we increase our speed.

When we finally arrive back at the Unit, it is already time to start preparing for Friday commander check. We spread a clean blanket on our beds and put all our equipment on it. Our commander arrives and makes sure we have all our gear and that it is clean and ready for use. Then he talks about the training we completed and what we can expect the following week. He discusses what we did OK (very little) and what we need to improve (a lot). We barely talk because we want the meeting to be over as quickly as possible.

Although it is still night, we are happy we can say, "Today we are going home for a weekend vacation."

Friday Check and Going Home

*Let not the foot of pride overtake me, and
let not hand of the wicked drive me away.*
(Psalms, Ch. 36, V. 12)

Morning brings Friday check with our commander, a discussion of the past week's operations and a briefing. Then, we get ready for weekend leave and going home. I am one among dozens of soldiers in a hurry to get home. I get a lift to the Sirkin junction, catch a ride to the next junction south. Thank the driver. At the pick-up point, there are soldiers from many IDF units. I look at them wearing black or red boots and hats colored brown, black, green or red. Each combination of colors identifies the wearer's rank and status. Black boots and hat means tanks, regular infantry and "Jobnicks" (non-combatants). Red hat and red boots are for paratrooper—serious fighters. We wear red boots and carry our hats in our pockets. Most of us prefer our field training uniforms. Only a few can recognize from our uniforms that we belong to the Unit.

I gaze at the soldiers around me. Many are much bigger and look much stronger than me. I look at the uniform that covers my aching muscles and that has been part of my identity for more than two years. I feel my painful shoulders and feet. I look around and compare. Stare at the soldiers who seem the strongest. I smile to myself. My haughty smile reflects the sense of pride flooding through my body. Warming my heart. I can't talk about this with the others but am aware of it in myself. The compelling desire to show off, even for one second, is too strong.

I feel my uniqueness and that of my fellow crew members in every cell of my body. We are not just "simple soldiers." I laugh. Only 15 hours earlier, I was practically unconscious and couldn't think any farther than properly placing one foot in front of the other. The warm sun is the extreme opposite of the freezing wind we left behind in the Hermon Mountains. The asphalt road invites me to walk comfortably, in marked contrast to the sharp basalt rocks that threatened to sprain my ankle at every step of the last march. The elastic bandages are still tied to my ankles to prevent any unexpected strain. Looking inside myself, a wave of pride fills me again.

A passenger car with a friendly driver stops nearby. I run over with other soldiers. Am the first to open the front door. "Where to?" I ask.

"Kastina."*

"Toda"—"Thank you," I say and get in the car happily. My rifle rests between my legs, the shoulder strap wrapped securely around my arm. I don't dare lose my rifle. Exchange a few words with the driver and the conversation fades off. I doze inside the warm vehicle.

Kastina.

The sun moves further west on its daily journey. I think of the chariot horses that pull the sun in Greece mythology. I'm feeling relieved without the heavy burden of dozens of kilograms of weight on my back. Barely notice the weight of my Uzi and the dirty clothing in my backpack. Don't feel the weight of my red boots with the crepe

---

* Kastina is a major junction on the road from the center of Israel to the South. From there, the main road heads to Beer Sheva and the secondary road leads to Ashkelon, the last city on the way to the Sinai Peninsula.

soles either. I am in harmony with my surroundings and focused on my coming weekend freedom.

I feel happy that I can show my worried mom that I am whole and healthy. She will be very pleased to see "her" soldier is healthy. I'm also proud that I can greet my dad as a soldier from the Unit. That is the best gift I can give him. His relentless demands I internalized being his son subside for a few moments as the sun warms our path south. Feel thankful for surviving the demands of the Unit training. Grateful. Get another ride. Near the Beit Kama junction, I see the memorial monument and remember the Azimuth navigation exercise we did ages ago. Get out at my kibbutz entrance, thank the driver and walk up hill towards the eucalyptus, pine and tamarisk trees. The pine and eucalyptus smells engulf me.

Enter my parents' home.

"Higati"—"I have arrived," I announce.

Mom and Dad reply and take my dirty clothing and get it ready to hand over to the kibbutz laundry. "Soldiers laundry" is done by a kibbutz worker in the central laundry.

Mom says, "You are thin."

In response, I inflate my belly and reply, "See how fat I am."

She laughs. My brother, Assaf, listens. My parents ask about my trip from the base and I tell them about the different rides. My parents have purchased a chicken to be fried in a pan. They also have brought vegetables and eggs for the omelet that I like to prepare for the family on Saturday morning.

Mom hands me three pairs of new socks. "To make it easier during your walks." I smile at her and stop myself before saying "they aren't really needed."

I exit the house and walk towards the dining hall. Along the way I meet people also walking the kibbutz trails who ask me how I am doing? At times a short conversation ensues. It is clear to me that my kibbutz friends love me and are proud of me. I feel content. At the same time, I know that I can't talk with them about most aspects of my military training. The warm sensations of family and friends leave me feeling alone with my army experiences that I cannot share. This aspect of loneliness has been part of me since enlisting. The demand "not to talk" creates an impenetrable barrier between me and others.

Yom Kippur—Eve of War

And thou saidst in thy heart: 'I will ascend into heaven,
above the stars of God will I exalt my throne.

(Isaiah, Ch. 14, V. 13)

On a Tishrei morning,* before the eve of Yom Kippur, 1973, we descend from the Hermon Mountain after completing a successful mission. My gear is piled in the truck and my Uzi rifle swings on my shoulder.

Although we are below the lower cable car station, the clouds are still below us. The first light revealed the striking beauty of the Hule Valley and the Manara Range to our west. In the local mountain valley, our cook Nathan is waiting for us. Partly limping, I walk down the asphalt road towards the valley. As we approach, we smell the wonderful food, including fresh rolls and a chocolate drink, made especially to indulge us "special soldiers."

Nathan has prepared a fresh salad and omelets. Amnon, with his usual wry humor asks, "Nathan, how do you manage to fry the omelets making them cold?"

I could not refrain and said to Amnon: "What do you want? The omelet is OK."

Amnon insists that the omelet is "not like the ones at home." We are starving and gobble down the food. We climb back on the truck and head south on the road toward the Unit resting in the back on our IDF mattresses.

* The first month of the Hebrew religious calendar, usually coinciding with parts of September and October.

218

The commander pushed us to finish Friday chores quickly. As we rushed to complete our duties, we are told that lunch will be at 1:30 in Kerem Hataimanim with Menachem Digli, a past commander of the Unit who is now in charge of information gathering for the IDF military intelligence. As a soldier, aware that missteps during Friday check-ins can lead to an hour or more of base detention before being allowed to go home, I feel confident that this Friday check can be taken more lightly. After lunch, we will be able to go home with no further delays.

The warm weather of the Tishrei month embraces us and is a pleasant contrast to the freezing Hermon air. We are all filled with joy, knowing that we will shortly be leaving for home. The thought of a brief vacation from the daily pressure of our grueling training, make us all happy, and a special meal is a welcome break.

Dressed in clean uniforms, we sit around tables in Kerem Hataimanim and feast on steak, pickled cucumbers, olives, and Coca Cola. The pickled olives included Broken Syrian Olives, a popular brand. Because the Hebrew word for "broken" also means "screwed-up," Doron never misses an opportunity to joke about "the screwed up Syrians" and has all of us roaring with laughter.

I pour Coca Cola in my glass and squeeze in some lemon. The sweet, bitterly-cold, bubbly drink fills me with pleasure. The ache in my muscles subsides and I am grinning. Before our meeting with a living legend like Digli, I feel excited and complimented. We worked hard and are receiving "Kol Hakavod"— "all the honor." Expressions such as this, foster our self-esteem and sense of "being the best." Since enlisting, we've all been nurtured to have this sense of superiority. We were told that our training is the hardest and that everybody expects us to demonstrate extraordinary performance and capabilities. We were told that utmost effort was required every minute of our service.

When we occasionally encountered soldiers from other IDF units, we heard the excitement in their voices at meeting soldiers from "The Unit." They admired our navigation abilities, our marksmanship, and the precision with which we performed attack maneuvers. We received recognition from others that the soldiers who served in Matkal are the best.

And we believed them.

Once, Nocheke from Mishmar Hanegev came up to me while I was walking the path around the kibbutz and said, "I heard from Yoram that you are excelling in the Unit. Do you know that makes you one of the best soldiers in the world?"

I looked at him, surprised.

He continued, "Yes, the soldiers in Matkal are the best the IDF has to offer and Israel has the best army in the world. Since you belong to the Unit and excel in your training, that makes you one of the best soldiers in the world."

I was stunned. Embarrassed. I stuttered a few words and moved away.

Digli, slim, six-feet tall, with the aura of the son of a Greek god, smiles. He compliments us for a job well done and stresses the importance of our work. "You know, never have so many owed so much to so few," he says.

We breathe in the heady air at the summit. Immediately, with no pause, he continues, "The IDF is on high alert, but you have nothing to worry about."

Despite the tension and high alert, we are going home for a holiday weekend. We will head home as soon as this pleasant lunch is over.

During the following days and months, I recalled time and again his statement, "The IDF is on high alert, but you have nothing to worry about."

We have nothing to worry about.

At the time, we all continued to enjoy the tasty food and Digli's compliments, which gave us great satisfaction. We were honored by Digli's presence. There was something so relaxing during these early afternoon hours prior to the eve of Yom Kipur. All was calm and peaceful. None imagined that the pleasant atmosphere would fall victim to a bloody sword. We all grew up feeling the sense of invincibility that was created from the Six-Day-War. It seems that Moshe Dayan's saying: "Better to have Sharem Al Sheik with no peace than peace with no Sharem Al Sheik," engendered a false sense of security and unwarranted conceit. We felt that Israel was omnipotent and we, its best soldiers, were supermen.

The sense that this was just another pointless alert was proven false in less than 24 hours. At 10 in the morning of Shabat during Yom Kippur, I was urgently recalled to the Unit. The Yom Kippur war was beginning. Only then did we start piecing together information that we had gathered but not understood. Suddenly we remembered looking through binoculars at locations along the Hermon Mountains facing Syria, and seeing armored vehicles and soldiers amassing month after month. This was no routine training. A few weeks earlier looking through the binoculars Adi said: "How do the Syrians keep from tripping over each other. There are so many of them?"

The Yom Kippur War. The war that changed our world. The war that cost the lives of Yosi Ur from my kibbutz class and my friend Muli Degani from Neve Shaanan, who served with me in the Machanot Olim youth movement. And many others. The war that shattered our arrogance and left behind a bloody trail of dead and wounded. The war that resulted in deep investigation of many of our military leaders. The war that tore apart and destroyed our belief in our leaders and created so much anger and distrust of every leader since.

The IDF is on high alert, but you have nothing to worry about.

Yehuda Amichai, the poet, wrote, "Anytime there are three or four people in a room, one is always standing by the window."

And we, part of the IDF military intelligence, who were supposed to stand by the window and observe. We looked but did not see. We were too wrapped up in our own conceit. When I remember the transition from feelings of all-mighty omnipotence to the recognition that we did not see right what was right under our noses, I feel small and weak. I live with the frustration that accompanies thinking that you "know," yet are blind to reality, and question, "Did we learn?"

Behind Enemy Lines

For thy work shall be rewarded, saith the LORD;
and they shall come back from the land of the enemy.

(Jeremiah, Ch. 31, V. 16)

The war was in its first days. Most of my fellow crew members joined forces that were hastily assembled and sent to the battles in the south. Yadi and I were about to go with a unit that was supposed to operate behind the enemy lines, west of the Suez Canal. Attended the first briefing. After a few hours I came to understand that "younger" soldiers were being assigned to replace us. I argued with the commander, Amiram Levin, and we were in. Then out again. Yoram from my kibbutz encouraged me to demand to be in. I was uncomfortable but went anyway, secure in the knowledge that "there is nothing to lose." Went. In again.

We attend another briefing, this time with Giora Zorea. Amiram is eager. Giora is more balanced. Yadi and I are happy that Giora is in command. The combination of Giora and Amiram seems like a good one.

We start loading the jeeps. Some are equipped with machine guns, others with "Tolar"—"light cannons." I take charge of the "Mag," the machine gun, and triple the amount of ammunition we are carrying, taking along as much as possible. Check that the load is tight and steady. Find additional space for more ammunition.

The order comes to move out. We are given an additional briefing on what to do if we need to evacuate. The Sea Stallion helicopter, a heavy-lift transport, lands on the paved runway and we get inside.

With the jeeps strapped in place, the choppers take off, heading towards the Sinai.

We land on a paved runway in the wind-blown desert. One last check-up. Drink. Take off again heading westward. We fly low. I stand by the window but can't see much. The sun set a while ago. Cross over the front lines. Tension recedes, replaced by the calmness before the battle. All the energy of getting organized is transformed into focusing on accomplishing our mission. I visualize myself climbing out of the helicopter and taking position a few dozen yards south of the paved road. We land in a small dusty cloud. Get out of the helicopters quickly and the pilots take off eastward.

We count off and make sure that all six jeeps are here. Move northward. The night is clear and there is no real navigation challenge. Just go north. I hold the Mag steady so it will not rattle too much. We stop. Giora walks forward and looks through his binoculars. He identifies the post where we will take our position. Our jeeps park there. We all are facing the road. Trucks move from west to the front line near the Suez Canal, driving with no headlights. More trucks than you can count. When do we attack? Not yet. I rehearse shooting the Mag in my mind. Remind myself to take careful aim. The battle will probably last only a few minutes. Little chance to waste all the ammunition.

"Look how many trucks," I say to Yadi who nods. Think, *Nu. When are we starting?* Not yet.

More minutes pass.

Giora talks earnestly on the radio.

Arguing? Nu. Give an order. Let's shoot at them. Make them afraid. How long do we have to wait? We are ready. Want to shoot. Nnnuuuuu?

"Go back," Giora orders.

"What?"

"Go back. Now!"

"Let us shoot them," I say.

"Go back. Now!"

"Let us at least show them. We are already here and can create such havoc with all this ammunition!"

"Go back. Now!"

Giora moves southward. We can't grasp what's going on. Follow him.

What happened?

Moshe Dayan called Giora and ordered him to pull back.

I tell Yadi, "What a missed opportunity. We were so close and they didn't have the faintest idea that we were out there. A complete surprise. What an opportunity. Aaargh! What a blunder."

We move southward and prepare for the helicopter's landing. Amiram's Jeep moves away. Where to? Securing the landing site.

The first two Sea Stallions land one after the other. Four jeeps are driven inside and the helicopters take off. We are in the third helicopter. We secure our own jeep and then Amiram's "light canon" Tolar Jeep, and the helicopter takes off. The gunner from the last Jeep is already dozing off, his bushy head resting against the canon trigger. I look at him moving with the bumpy flight. The hair on his head continues to "caress" the trigger. I look out the window into the darkness. The illumination from thousands of tracer bullets passing near us lights a path across the dark sky. Two minutes of intense shooting and we cross the front line. The canon gunner continues to doze, gently swaying back and forth over the gun trigger. A few more minutes and we land.

I climb out and stand on the paved runway next to the pilot. He holds a flashlight and shows me the numerous bullet holes in the rotor wings. "That was close" he mutters.

The canon gunner gets off the helicopter and says to me, "Oops, the canon was loaded and not locked."

I imagine what could have happened if he had accidently pushed the trigger. A cannon shell launched towards the helicopter cockpit. I tremble. Take a deep breath. Another. Think, *Thanks for keeping me alive so far.*

Loud thunders from shells exploding to the west. We are safe. For now.

Bombarding Tel Shams

And I will cause to rain upon him, and upon his bands…
an overflowing shower, and great hailstones, fire, and brimstone.
(Ezekiel Ch. 38, V. 22)

During the Yom Kippur War, Israel conquered Syrian land, getting close to Damascus, and positioned forces on several lava hills on the Golan Heights overlooking the Syrian plateau. On one of our missions, we were posted at Tel Shams,* one of the volcanic mounds, for a few days. The Tel was scarred by many cemented trenches, and a large bunker stood at the middle, covered by several protecting basalt stone layers. In the main bunker, there was room for food, ammunition storage and sleeping mattresses placed on double bunk beds. All of the tunnels led to the main bunker. The trenches had been dug deeper than the height of an average-sized man. When there was an alarm that the shelling was about to commence, we would jump into the nearest trench and run to the bunker. Along the trenches, a few defensive forward posts had been built, allowing a few soldiers to serve as lookouts and warn people in the main bunker of impending attacks.

On the first day, as we familiarized ourselves with the surrounding area, a 30-year-old reserves soldier advised us, "Tel Shams holds no interest for the Syrians. You can see that other hills are getting heavy shelling while here it is very quiet."

* "Tel" is Hebrew for "hill" or "mound." "Shams" comes from Aramaic for "sun." Tel Shams means "mound of sun." It was a large hill strategically positioned at the southeast corner of the land "taken" from the Syrians during the Yom Kippur War.

We spent part of the day in the forward posts and the rest of the time in the bunker. We liked to sit outside on the Tel's slopes looking around and enjoying the pleasant weather.

After two days of settling in, we felt comfortable enough to take sunbaths on the slope. We looked out at the other Tels, at the Syrian countryside and to the far away Hermon Mountain. Tel Shams was at a very high altitude, and we could see supply trucks moving on our side as well as on the Syrian side. During the night, we watched the lights from their vehicles moving on the roads. A few fainter lights signaled the locations of their soldiers. The nights were warm and peaceful. On our post, we felt as if we were on an island of calm in the middle of the storm. Syrian shells fell on other Israeli posts, not ours. We couldn't engage in physical training exercises since we were continually on alert, always ready to take cover in the bunker. I had the feeling that all the warnings didn't really apply to us. But, we had to observe the rules. Maybe? After the Yom Kippur War, all belief in "prophecies" evaporated.*

On one morning, a command car stopped at the Tel's entrance and Dror, a soldier from a younger crew, got out. He came into the bunker, unloaded his gear and later joined us on the slope. We were in the middle of cooking lunch. Combat rations were significantly upgraded after the war, and Luf, the ground meat that was barely edible, had been replaced by a "well-cooked" chicken. It had been cooked so long that the bones could be eaten without chewing. Tasted as you would expect. Yet, compared with Luf, it was fine dining.

We placed the chicken in a pan with olive oil and onions and fried it up with a gas torch. We cut up cucumbers, tomatoes, lemon

* Because the IDF intelligence had predicted "no war," the Israeli
  leadership failed to anticipate the conflict, so after the Yom Kippur War,
  no one took "crystal gazing" or intelligence forecasts seriously anymore.

and peppers, and doused them with olive oil to create a fresh tasty salad. Within a few minutes, we would be feasting like kings. We looked out towards Syria and the other Tels where IDF soldiers were subjected to heavy shelling on a daily basis. The weather was calm and comfortable as we spread out our plates and silverware. The feast was about to start. I lay on my back, resting. Eyes closed, I started hearing my favorite Beatles' song, "Because," inside my head. Enchanting melody. I am calm and relaxed. I open my eyes and look out at the Syrian Tels. Everywhere the front lines are quiet.

Dror comes in from the trenches and asks us, "What's up?"

We say, "Come, join us and eat. It's real fun here."

Dror sits down, takes a plate and start eating with us.

"Take off your helmet and shirt," I say.

"Don't you worry about getting hit?" he asked.

"No."

With some hesitation, Dror takes off his helmet, settles in and enjoys his food. As we eat, we all get more relaxed. Dror unbuttons his shirt. On the Tel north of us we hear a few explosions from shells. We slowly finish our tea at the end of our meal.

Dror asks, "How can you sit here so calmly and not be afraid that a shell will hit?"

I give him our routine reply, "The Syrian have no interest in us so they never shell this Tel."

Suddenly, everything changes.

There is a loud whistle, followed by the explosion of shells.

The party is over. We instantly turn into Olympic runners. Within seconds, we each jump into the trenches and run to the bunker, with Dror being the gold medalist. Fear ignites our responses like rocket engines. All of the post soldiers are gathered in the main bunker. The bombardment continues intermittently at first, then steadily as more shells hit the Tel and we finally stop counting. We look at each other and see that every man is dressed as a soldier again, wearing his uniform and helmet.

"What happened to the continuous picnic on this Tel?" asks Dror.

"The Syrian must have heard that you were coming because up to now everything has been quiet," I say.

The shelling continued at random intervals. The concussion from the blasts caused sand and small basalt stones to drop from the ceiling onto the occupants of the main bunker. We felt protected buried deep beneath the earth's surface. We called the Unit headquarters and reported.

"Pack up and get ready," headquarters told us. "A command car will pick you up tonight and bring you back to the Unit."

We packed our sleeping bags and other gear in our special bags called "chimidans." When the night came, the bombardment declined, then ceased altogether just past midnight. At 1:00 a.m., the command car arrived. We loaded up and set off for the Unit. The trip was very slow because the driver tried to avoid the hundreds of pot holes in the road, caused by the intense shelling. He kept shifting between first and second gear, applied the brakes, then drove forward in first gear, inching along. The road had deep crevices from tanks crossing it. Wide holes. The driver continued to maneuver between the obstacles. I dozed off and woke up because the command car was shaking.

As we cross the old border from the 1967 war, the road improves significantly and the driver shifts into higher gears. We are heading home and I inhale the cold, clean night air deeply. Each man dozes in his corner of the car. We are heading west, home to our land, to Israel.

During that weekend, the Syrians blasted Tel Shams with enough ammo to more than compensate for the lack of shelling of the previous weeks. From the weekend newspapers, we learned that the Tel that had been "of no interest to the Syrians," had become a primary target and was completely destroyed.

Hagai Verner

Yadi and I train with other crews and our team goes to the Hermon without us. We receive a message to go to Rambam Hospital in Haifa to be beside Hagai who got critically wounded in the same accursed Hermon where Shai died.

We arrived. Hagai just got out of nine hours of surgery performed by one of the best brain surgeons in Israel. The bullet of the Syrian sniper entered through his eye and shredded part of his brain. He is alive but in a coma, and nobody knows his fate. Sitting on a bench in the hospital, Yadi and I promise each other that we will do whatever it takes for neither of us to be in such a state. Miri, Hagai's girlfriend, waits. She continues to wait for six weeks until Hagai moves his finger. How much joy moving a finger can cause? We all pray that the rest of his body will follow. Hagai continues to improve and recovers during many long months in Levinstein center in the sapping, frustrating rehabilitation. Marching in thick tar. We visit whenever we can. Miri is beside him. Supports him throughout the long rehabilitation trail and the ensuing years. Hagai and Miri are heroes without having a choice. A source of inspiration.

From Rambam we went to see Adi in Zefat Hospital. Another sniper's bullet "carved" a hole in his ear. Adi and Hagai were together in the evacuation zone close to the Hermon summit. You know how it is so high up at 9,000 feet—little oxygen."

Yadi and I nodded.

Adi told us, "After I was hit, Hagai bandaged my ear and said to me, 'Don't worry, I glued your ear.' Then he returned to the battle.

Suddenly, his head dropped forward, hitting his rifle. The two doctors who were near him in the crater, turned him on his back, took a look, and declared him dead. When they turned him face down, I thought to myself, *We lost Shai and now Hagai is gone.*

A few minutes later, I noticed that Hagai is contracting his left hand fingers. I shouted at the doctors, but they couldn't hear me because of the wailing wind and the shelling. I got up, grabbed one of the doctor's hand and shouted in his ear that Hagai had moved. The doctor looked at me in disbelief. We both approached Hagai and his fingers contracted again. The doctor started to vigorously treat Hagai, preparing him for evacuation with the approaching helicopter.

The helicopter landed and we loaded him inside and it lifted off, heading towards Haifa."

What would have happened if Adi had been wounded more seriously and had not seen Hagai?

The distance between life and death so short.

Hagai continues. His youthful vigor diminished. Yet continues. Married. Two smart beautiful girls and a granddaughter he is very proud of. The life force pushes each one of us. Hagai continues. Despite the limitations and difficulties, he struggles and continues. Day after day for four decades, until today.

Encountering the Commander

Show me now Thy ways, that I may know Thee,
to the end that I may find grace in Thy sight.

(Exodus, Ch. 33, V. 13)

At a gathering of the Unit's veterans a while ago, I met Yadi. We entered the high ceiling hangar loaded with gymnastic training equipment and started talking.

After a minute, surprisingly, I feel someone massaging my shoulders. A hoarse, familiar voice says: "I was told, 'Go on this line and you will see Ouri.'"

I turn around. See the commander. He continues, "Long time since I saw you. How are you?"

A flood of feelings and thoughts surge through me at light speed. Colliding. I am stunned.

I can feel the protective armor starting to assemble. Sealing any crack of weakness. My diaphragm contracts and breathing accelerates. Sweat in my armpits. Facial expression frozen. I am burning and on alert. Forty years passed and I respond instinctively. Pavlovian reaction? I breathe.

I can recognize the sharp facial lines that have slackened over the years and are now covered with a little fat. The tense physical posture relaxed a bit. Small paunch. The tiny smile that transformed from critical to apologetic. All the same. All so different.

His eyes.

The same brown eyes, but the expression…softer, warm, asking for intimacy and friendship. Almost begging.

The acute difference is stunning. I am aware of the rising tension in my body.

We talk. Exchange information about our children, wives and homes. It feels very strange to have this conversation. Almost calm while in the background thoughts and feelings are still colliding. Almost reach the surface. I stop the thoughts before uttering them. An emotional storm accompanies the "civil" conversation.

I attempt to drop the armor. To become softer. My body cells rebel. Not ready yet. Remind myself that forty years have passed. But my body has its own timeline. Relive the fear of being dismissed. Feel again the pain of the unjustified termination of Avner and Dany. The scorching loss of Shai in the Hermon.

The commander faces me. And I am flooded with experiences from forty years ago.

He was only one year older than us. One year.

I am aware to the gap between the stressful, demanding atmosphere then to the soft calm surrounding us now.

Why am I so tense?

Remembering.

Before we arrived at the Unit, many were dismissed. Thirty-two of us started, only sixteen made it. Throughout the long months of extreme training, the relentless demands were accompanied by the threat of dismissal. Decades later I was asked if I did not know that I was a "good soldier"? I responded that I knew I was good, yet afraid of being dismissed. Other good soldiers in our crew said that they felt the threat of being discharged throughout training, too. Forcefully. While meeting high standards of performance. Soldiers in other crews did not feel it so powerfully and not for such an extended period of time. How does a person behave when he is told that he is "the best" while the guillotine blade of dismissal hangs over his head?

I recall a significant moment.

During the spring of 1974, as we returned from one of our afternoon training sessions, we drove on the narrow Golan Heights roads. The weather was lovely for a change, and we opened the truck cover to expose our bodies to the refreshing wind. We had finished another grueling exercise. The light was clear but not blinding. The caressing wind effortlessly covered us all with good spirits. The sun went down slowly.

Within a few minutes we notice that the truck accelerates and takes each curve in the road with increased speed. We are thrown from one side of the truck to the other. When more curves are maneuvered at break neck speed, we knock on the driver's cab back window signaling to slow down. The commander glances at us and motions the driver to accelerate further.

Our fear increases with each passing curve. We try again and again to get the commander and the driver to slow down, but to no avail. Panicking, we grab hold of anything to prevent us being flung from side to side. I start pounding with my fist on the tin roof of the cabin. The truck stops abruptly with screeching tires. As the horror abates, we all take a deep breath of relief.

The cabin door opens and the commander climbs out and walks to the rear of the truck. "Who pounded?" he asks.

"I did," I reply.

"Get off the truck," he commands.

I get off. The commander starts marching towards the cabin. My friends get off the truck. He stops. All the crew soldiers are off the truck.

"Get back," he orders.

No response.

"Get back on the truck!"

"All of us," Adi and Ovad say and the rest stare boldly at him.

Silence.

The sun continues to go down. We all stand at the back of the truck and the commander near the cabin. Another long silence. The driver watches, stares dumbly at the steering wheel.

There is a commander and there is a crew. And, an order disobeyed.

Silence.

Thoughts cease. A group against an individual. The Hermon landscape on our right and the Golan Heights stretched westward. Magnificent sunset.

"OK. Get up."

We all climb silently and sit on the bare benches. The driver engages the gears and starts moving at a reasonable speed.

At the temporary base, we eat supper in silence and get ready for the night exercise.

I am back in the present with the commander in the hangar at the Unit.

What happened with him during these years?

Where did the rift start? The breaking point?

Hermon.

Hermon with Shai. Frozen Shai.

The commander, a highly promising young man. The son of a university professor. Excellent athlete. Excelling soldier. Soaring on strong winds of support. He received the opportunity to command a crew and the crew excelled. He receives all the credit.

Then, all of a sudden, a shuttering break. His world collapsed. Broke into countless pieces.

He is alone. His Icarus ascent halted. His admirers turn their backs on him. He attempts to gather the broken pieces. Accompanied by the unforgiving guilt of Shai's death. Like a shadow stuck to his every step. Daily, weekly, monthly, and year after year. Decade after decade. Only Ben Ami and Ovad notice. Can see his pain. The rest of the crew members can't. Won't. They are indifferent, angry, unforgiving. Remember his harsh, demanding conduct towards us.

Did he relate to us in such a way due to ignorance? Was he too young? Was he so tough with us because he lacked knowledge? Many crew commanders in the Unit had harsh relationships with their men. What was so different about ours?

The operational results of our crew were superb. We reached levels of achievements that had never been accomplished before. Yet, we never told anyone that we were responsible for the navigation successes and performed the tasks on our own.

We just did it.

I remember going through those experiences detached. I did not connect to the anger. Looked at the commander's behavior through philosophical glasses. Justified his actions. Did not allow myself to feel. Attempted to calm my friends with Shai's assistance.

After Shai froze to death the commander fell apart. He attempted to get close to us but the burning pain enlarged the gap. Fed the growing distance between him and us.

Only decades later can I empathize with his meltdown. Deeply understand that he was only a boy. Like us. A human. See his on-going efforts to get closer to us rebuffed by our "cold shoulder." Continued intensification of social isolation.

With his last bit of energy, he attempts to climb, his wing's wax melting. He collapses inside.

We had ourselves.

He was left alone. Unconsoled. Ostracized.

Secret Mission on the Golan

Sun, stand thou still upon Gibeon;
and thou, Moon, in the valley of Ayalon.
(Joshua Ch. 10, V. 12)

And so we ordered darkness, "Stand still!"

On one of our missions after the Yom Kippur War, we were required to walk through the darkest part of the night, perform our mission and return before sunrise.

Walking back. A line of soldiers. That part of the mission is blurred in my mind. All of us are focused on crossing the border back into Israel before first light. The weight we carry is burdensome. All my muscles are tight. We will complete the march on a Tel, where the truck is waiting for us. From there, the truck will travel south while we sleep. Cross the last goat trails past the junction, curve right with the terrace. My shoe laces are tied tight to support my ankles and prevent sprains. The shadow of the Tel rises in the distance. Getting close. We are on a wide basalt road moving uphill. Breathe heavily on the last ascent of this night. Almost there. Get closer. See the soldier ahead of me stop. Hear voices. Another step forward. Surprised to see Raphael Eitan, the general in charge of northern military operations. He shakes my hand and says, "Kol Hakavod" (All my respect). I reply, "Toda" (Thank you), and keep walking. I feel immensely complimented. A sense of pride fills me and helps ease the pain of my aching exhausted body. Climb to the

Tel's highest point, move further west, unload our gear, and get into the back of the truck for a restful trip south.

I remember how we began training for this kind of mission over a year ago. When we set out, we were told that we have four hours to get there, four hours to do our job and four hours to march back. We began training for these missions. The first attempt involved work only, during the day, and took us 16 hours. Consistent with the spirit of the Unit, our performance was evaluated and critiqued to improve efficiency and shorten work time. Before each exercise, we were briefed, went out into the field during the morning and started performing the complex mission. Ehud Barak, the Unit commander, watched us and took notes of us while working.

The sun goes down. We stand around Ehud. One by one, he reads off his written comments. There are 193 altogether. His comments include, "Ouri, you did not tie the carrier strap tight enough and thus had to retie it five minutes later." And, "You left the carrier and didn't pick up the small bag. After a minute you came back for the small bag." We listen and digest. Pack our gear and return to the Unit.

We arrange the gear for the next exercise. Briefing. Out in the early morning. Ehud watches and writes down comments. Work with extra enthusiasm. The sun sets. We cut the time to complete our task from 16 to 12 hours. Ehud reads his comments—"only" 103 this time. Pack the gear and return to base to get ready for tomorrow's exercise.

Next morning, we get briefed, go out into the field and perform the task again. This time we finish within eight hours. Ehud reads fewer than 70 comments. Our goal is to solicit fewer than eight comments and complete our task in under four hours. Achieving this seems impossible. Beyond our reach. Most crew members discuss the comments. Suggest creative, out-of-the-box ideas for improving performances.

After weeks of intense training, we load the gear, get briefed, head out early in the morning and work. This time we managed to complete our task in less than six hours. Only 42 comments. Despite the improvement, we are still far from achieving our four-hour goal. We pack up the gear and go back to base.

Practice, practice, practice. More weeks pass. Practice for months. Practice. We are close to hitting the four-and-a-half-hour mark with fewer than 20 comments.

More practice. Less than 15 comments. The pressure to complete the task in less than four hours is increasing. Our motivation increases minute by minute and we manage to shave extra time from the work performance. The complex task is performed without a word spoken. We communicate among ourselves by extending a hand, blinking an eye, or simply nodding. No words. We are speaking less and less. Like cloistered monks, the "vow of silence" becomes a part of us and stays with us for years after leaving the service. We are each part of a group where every man knows his job and performs it with precision, in coordination with the other members. The operational crew becomes a sophisticated, efficient human machine. A delay caused by one member is shortened by another and the ones nearby. A living breathing organism. Operating in sync. Each focused on reducing time and errors. We are a group of "engineers," fine-tuning the workings of a "living machine." Each day. We are the machine parts. Improving and being improved. Work until the mission is accomplished flawlessly.

Yadi and I work closely together. A couple with a common mission. Yadi has exceptional technical capabilities. His fertile brain and competent hands unravel each technical challenge. I am his "Number Two." Our task takes the longest. We spend hours weighing options for completing our mission. Examine each aspect of an action diligently. What unnecessary steps can we take out of the process? How can we eliminate any possibility of error? In our discussions, we toss around unexpected problems that might arise during the process. Address the problem and implement a solution.

I wake up in the morning with an idea of how to cut another minute from the process and we go ahead and try my solution. If it works, we integrate it into the industrial assembly-line process. If it doesn't, try another solution. We are very focused on producing quality results. Hundreds of actions required to complete the mission. Every movement of hands, feet, fingers, tools is assessed. How do we eliminate extraneous actions? Can we replace this hammer with a special hammer? Should we build a device that will speed up a series of actions? What will it weigh? How can we divide the work better? Can another person assist in some part of the process? We continuously think about shortening the time and increasing the quality of the product. No room for error.

Any resource we want is made available to us. No limit. The only limit is our thinking and getting stuck in our ruts. With Yadi's creativity we experienced breakthroughs on a daily basis. We lay a foundation for a certain frame of mind and built on it again and again. Time was limited. We had to condense and cram and transform. The creative process fired within the furnace of limited time becomes our daily routine.

We felt like we were the first among "the chosen." Yadi was a born leader. As his Number Two, I felt proud and happy to work with him. It still amazes me how little we spoke and how much we accomplished. Yadi cracked the technological challenges and together we engineered the solution. A joyful, creative undertaking. A gratifying process that drew us back to the Unit every Sunday after weekends off.

Later on, in his fifties, Yadi had an operation. A benign tumor at his brain stem. After another year, another tumor was found in his thyroid. Operated on again. As Number One, Yadi carried dangerous gear. Week after week. Two or three times a week. For more than a year. Today he lives on a farm in Nahalal. Married with two daughters and a son, and is a grandfather. After four decades, the effects of his years of military service are showing on his body. As soldiers we had no awareness of our fragility. We, who could perform any impossible task, are "fragile"?

That word did not exist in our vocabulary. We despised fragility and weakness. But four decades later, we realize: we are flesh and blood.

Part of our training was grueling, forced marches, preparing ourselves to walk to the mission site, perform the assigned task and return within the allotted time frame—in one night. Training and more training. During the months of training, I start to understand that we are building the physical and mental stamina that will enable us to meet the required standards for performing the mission. When I told Yoram about the grueling marches in the Hermon Mountains and the Golan Heights, he said, "It is not that the tasks are getting any easier. They are actually getting harder. The difference is that you are getting used to exerting the extra effort."

Training. We constantly challenge our bodies. All our muscles are tight and continuously hurting. Marching. Through fog, rain, and freezing wind. The rain pelts my right cheek and ear mercilessly. I console myself with the thought that on the way back, the rain will be pelting my left side.

I challenge the obstacles and aches that accompany me during the long nights. I mock my suffering. We can see our progress from exercise to exercise. We improve with each new training. We are joyful each time we transform the impossible into the possible. We rise to the challenge.

A few years ago, I read *Gates of Fire* by Steven Pressfield about the lives of the ancient Spartans. I was excited by the book and bought several copies to hand out to my friends. I read the book several times. Suppressed memories from my years of military service began to resurface.

The book is told through the eyes of an aide to a Spartan warrior, from the beginning of his life until after the Battle of Thermopylae.

In 480 BC, a Persian army of two million men, under the command of King Xerxes, marched to conquer Greece and crush the newly born democracy. To invade Europe and conquer the whole known world. An elite force of 300 Spartan warriors, under the leadership of Leonidas, the Spartan king, marched from Sparta to the Passage of Thermopylae to stop the advance of the Persian Army. They and their allies, the Thespians, fought and blocked the Persians at the narrow Thermopylae Pass for a week. All of the Spartan warriors were killed. No man survived. Their courageous stand inspired the other Greeks who, during the following winter and spring, defeated the Persians in Salamis and Plataea and thus kept the buds of freedom and democracy alive.

My interest in this historical event caused my family to journey to Greece. We rented a car and drove to the Thermophile Pass, an hour and a half trip on the autostrada north of Athens. At Thermopylae, on the Memorial to Leonidas, his words to the Persian King's emissary are inscribed. The Persian king demanded that the Spartans surrender their weapons to the Persians.

Leonidas replied, "Mulen Labe!" Meaning: "Come and get them!'

My Dad also challenged his oppressors, replying "Mulen Labe!" to their threats in his language.

His survival was his way of saying "Mulen Labe!" to his oppressors and the murderers of his family. His people. In the Krakow Ghetto, in the camps of Auschwitz, Birkenau, Buna Monowitz, Buchenwald. The ability to fight was within him. In his blood. The challenge chose him. No Spartan training. How did Dad confront the Nazi terror with no Spartan warrior to mentor him? The frustration and choked cries of rebellion were already internalized in his soul. Later, the frustration and choked-back screams resonated in my soul as well and were nourished by the relentless, grueling training.

When I was growing up, my father told me a few stories that stayed with me through the long marches and along the IDF service. His story of the Death March from Buna Manoviz to Gliviz through deep snow and freezing cold. It began January 17 and ended the night of January 22, 1945. The night of the January 21 was especially cold. There was no place to get warm. As night fell, my father found a bare wooden door and covered himself with it. A little after midnight the freezing temperatures woke him. He attempted to wake up the people around him, shouting at them, "Shteit off. Shteit off. Eir vat bald zein far froiren."—"Wake up, wake up. You are going to freeze to death if you don't move."

A few woke up and joined him stepping in place to warm up. Other remained unmoving and perished. During that night thousands froze to death. The next morning, which was the sixth day of the Death March, my father got up with a piercing pain in his hip. "I could not take even one step," he told me. The ball of the hip bone ground against the hip socket "like sand paper." When he asked to be left alone, his friends didn't listen and insisted on carrying him. He argued, "I will cause you to fall behind." They were adamant. With their last ounce of energy, they put his arms over their shoulders, lifted him up and began to limp forward. Supported by his two friends, he moved in the line with the rest of the "dead." He continued to beg them to abandon him and they refused, insisting on helping him. After a few hundred yards, his joints warmed and he was able to limp forward on his own. Marching as part of the endless column of human suffering. So they marched, supporting each other, until they arrived that night at the train in Gliviz that was to take them to Buchenwald. When he told me this story, I felt immense gratitude to his friends.

How did he do it?

The hopeless atmosphere, scant odds of surviving, inability to control your fate, the deep frustration, the desire to defeat a cruel enemy. All of these feelings are part of me. I tapped into them during my training with the Unit. I am not sure that I made a deliberate choice. The unconscious desire to ensure that my family and friends would never be in a similar situation predominated during the grueling months of training. More than me choosing this road, the road chose me. I recall after finishing very difficult marches asking my father to tell me about the Death March. Time and again, everything fell into perspective. Each time I understood that no matter what hardships we endured, we would never come close to measuring up to the strength and courage of my dad and his friends. The survivors of the Death March. And it was all made so much more difficult for them by not knowing if and when their struggles would ever end.

During the arduous marches that were part of our training, we always knew how long the exercise would last. And we exerted ourselves, knowing that it would end. And that, when the time was over, we could rest.

How Dad got up day after day facing another day of intense physical labor with little food and weakening muscles? He never had the comfort of knowing "within 48 hours my situation will be so much better." Not even 48 days or 48 months. That we knew our exhausting task would be over in a matter of a few days or hours was highly comforting. Even exercises that attempt to simulate endless efforts would eventually be over. If not this week, then a few days later. And at the end of the exercise, we knew we would return to the base, shower, eat a good meal and go home to rest. We were not trapped for eternity in a place of freezing cold, constant hunger, continuous threats, and incessant reminders of our weakness, surrounded by electric fences and vicious guard dogs ready to tear our throats out.

We trained. Marched, accomplished our task and came back. Every minute was counted. I found myself calculating the remaining minutes of the march relative to the plan. I created an internal timekeeper. A silent companion entity that watched, counted, accompanied and encouraged while walking. The dimension of time became a component of my reality and an integral part of space. No need for a watch.

The awareness of time had us placing bets. As we start our journey heading towards the base we would wager. "Another three and twenty-four." First bet. "Three nineteen." Second bet. The time that passes before we return to the base. During the journey we bet on the time to arrive at milestones along the way—Afula, Megido, Beit Lid. The road is sliced into minutes. At each point the winner is heralded. His prize? The acknowledgment by the rest of us of the winner's statement "I am right! Of course!" The losers nod smiling, gracious in defeat.

During the trips, some soldiers sit in the back, watching the passing road. Sitting in the back of the truck allows you to enjoy the scenery but also has a few disadvantages. The back of the truck is bouncy and some potholes cause a painful jolt. Those sitting in the back also can't help inhaling dust and exhaust smoke and are covered with an extra layer of dirt. During rainy days, the soldiers riding in back get wet. The soldiers in the front experience a less bumpy ride and better air quality but can't see the road. I usually prefer to sit or lie in the front and take a good nap. Usually we lay on IDF mattresses dozing until we reach our destination.

We worked and worked to shave every possible second from our task. The number of comments is getting smaller and we are getting close to hitting our target time. We continue to train waiting for the right moment. "In a week," we are told. Prepare our gear and study the trail. Check and re-check orders. Briefing and more briefings. Continue to train.

The day arrives. We travel north. Unload the gear and get dressed. I am not sure about wearing my wind breaker coat. Although it is cold and the wind is blowing, I know that after a few minutes of walking, I will be "boiling" hot. Decide to take off my warm shirt and put on the wind breaker behind my army belt. We pick up the heavy carriers and start walking. We are all breathing heavily as we descend the Tel. As we reach the plateau, our pace picks up. To cool off, I move my coat tails behind the army belt. Think: *Lucky me to leave the shirt off.* We are a heavily loaded mule team marching energetically, sweating profusely and breathing heavily.

I have a problem taking a deep breath. I can't fully fill my lungs. I am familiar with the discomforting sensation of not being able to take a deep breath and don't stress out about it. I know that at some point I will be able to breathe deeply again and am satisfied for now that I am inhaling a sufficient amount of air. I know that because this sensation of not having enough air has accompanied me through many months of training. As we move forward, my breathing is better and more balanced although I can't yet take a deep breath. We

march fast along the dark trail. I tell myself we will arrive at the target two minutes ahead of schedule. We reach the target and do the
job. Organize our gear and walk back. Despite our exhaustion, we
hurry back, drawn to home like a powerful magnet.

Saving an Arab Truck Driver

Heading home early on a Friday. Barely noon. Get close to the Plugot Junction. My ride stops before the junction. A crowd has formed on the road around a lime truck turned over on its side, facing southeast. It looks like the truck flipped over when it was attempting to turn left. I see the lime spilling onto the asphalt. I hear shouts, "The driver is trapped!"

I sling my weapon behind my back and hurry over to the truck cabin. I see that I am alone. The crowd keeps its distance on the roadside. They watch intently but don't come forward. A circle of people, black asphalt, and an overturned truck at the middle. A wave of fear floods me. I continue to advance towards the truck cabin while fighting back my rising fear. I tamp it back down firmly. Get closer to the cabin and farther from the crowd. Alone. Three feet to the cabin. The fear evaporates. I am on a mission. I see the driver. Half of his body is out the window and half is trapped under the truck. Crushed. Arab. Certainly Arab. The lime continues to leak out on the asphalt.

"Jack," I shout. "Bring me a jack!"

Within seconds, three men are beside me and one hands me a jack.

"Help me to place the jack here," I command and the men assist me.

"We'll get you out," I tell the Arab driver.

I am not sure he understands me. He moans in pain. The jack starts lifting the truck up. Another moan. The truck cabin is a few inches off the ground. I see that the driver's body is freed from the weight of the truck's cabin. Another man and I start gently pulling the driver out. I place my hand under his head for support while pulling him. He cries out and moans. We continue to pull slowly. I talk to him calmly, "We will take care of you. You are OK." The driver's voice is quieting slowly.

Red Cross paramedics arrive. The ambulance stands nearby. They bring a stretcher and transfer the driver onto the stretcher, just as we had been taught to do. Very professionals.

Small lime stains are on my IDF shirt. I go back to my ride and ask to continue south. "No reason for us to stay here any longer," I say. "They will take care of him."

I think back to the moment when I approached the wounded Arab driver who was trapped under the truck, alone. At that moment I felt piercing fear. Fear of what? I have no answer and do not continue to explore.

The sun is warm and comforting. I relax in the car seat. Arrive at Mishmar Hanegev. Thank the driver for the ride. Thankful for the early outing. Think about my classmates whom I will meet again soon. Tell myself to enjoy the respectful looks I will get as I enter the dining room. Walk up the entrance road to the kibbutz. The scents of the eucalyptus and pine trees welcome me.

Expulsion from the Unit

Therefore, God sent him forth from the garden of Eden,
to till the ground from whence he was taken.

(Genesis, Ch. 3., V. 23)

Only a few months are left until our official dismissal from the IDF, which was delayed because of the Yom Kippur War. The period following the war was very busy for us as the lead operational crew. We spent many days and nights in the Golan Heights, enduring the freezing winter and the hot summer that followed. The death of Shai was always present in our minds. During our mission training and the actual operations, another commander joined us. Yadi and I performed several missions with other crews, which gave us extra breathing space that our friends did not have.

On one of the mornings in the Golan, the commander told me to bring him a map. I muttered, "I will bring it in a minute. I need to finish what I'm doing first."

He was irritated, probably because of the tone of my response, and said, "Do it now."

I tried to evade him, which made him even madder.

After a short argument he told me, "Pack your gear and go back to the Unit. Tell Giora that I expelled you."

I felt relief but also frustrated. I thought it was an overreaction to "throw" me out of the Unit at this stage with all of my accumulated knowledge and experience. I hitchhiked from the Golan back to the Unit carrying my chimidan (IDF bag) with me. I felt a fatalistic sort of surrender. *What will be will be*, I thought to myself. But I felt discomfort, shame and embarrassment at the thought of facing Giora, the Unit commander. A few months earlier, Giora offered a number of times to send me for officer training. He was with our crew a little while ago on a mission and consulted with me. What will I say to him? That the commander "threw" me out? During the long hours traveling south back to the Unit I envisioned myself standing before Giora and telling him that my commander "threw" me out. This picture tormented me.

I identified myself completely as part of the Unit. The concept of no longer sharing this existence was difficult for me to fathom. I felt paralyzed confronting such a reality. I could not envision what would happen afterward. I could not imagine where my life would lead after the intense activity that was cut short this morning. I was not worried about where I would go next. I was concerned about facing Giora. I did not want to stand in front of him, but I had no choice.

I arrived at the gate and walked towards the Unit headquarters. I stopped by my room to place the chimidan on my bed and went to the offices. To Giora's office.

"Is Giora here?" I asked his secretary.

"Yes. Go right in."

Giora sat behind his desk. when he looked up, I said, "Giora, my commander expelled me from the Unit."

The moment I dreaded, had passed. The next moment, surprised me.

"Well, Ouri, I want you to take command of Galili's crew. He is wounded. OK?"

For a moment, I was speechless. "OK…" I muttered. "What do I need to do?"

"Prepare a work plan for next week's training in the Golan. When you're done, come back and sit with me."

"OK. When should I come back here with the plan?"

"In three hours."

"OK."

I left his office, feeling relieved and joyful.

Crew Commander

Who so loveth wisdom rejoiceth his father.
(Proverbs, Ch. 29, V. 3)

With mixed feelings I took over the command of the crew. I enjoyed the work and the opportunity to apply my knowledge and experience to an operational crew. But it was difficult for me to tell others that I was in charge. At one point, I accompanied the crew to a firing range for target practice with pistols. The instructor looked around for the commander. When he realized that it was me, he asked, "Why didn't you say you are in charge?"

I felt very good about my ability to lead but very uncomfortable with the title of "commander."

I worried that my soldiers would find me threatening and intimidating, as I had felt many times about my commander. I was concerned that I would be too arrogant and treat the soldiers in a condescending manner. I was afraid that I would demand too much from my men and ask them for more than I was willing to do myself. I realized that it is nearly impossible to be loved as a commander because of the extreme pressures put on the crew, but I wanted to be respected for my professionalism. My own experiences of not trusting the judgment of my commander in an operational situation was still fresh in my mind. I was alert for any signs that the soldiers didn't trust me as a commander. In addition, I did not want to prolong my military service, and I made an agreement with Giora that each month we would review whether I was still needed as the crew's commander. As long as that was the case, I would continue to serve.

I fell into a routine. Prepared a work plan, sat with Giora to review and approve it, then implement it. At the end of each week, we discussed what had been accomplished and consulted about the next week of training. Giora asked questions, I explained to him what I had planned, and he interrogated me about the details—the details I placed less emphasis on.

"When did you set the rendezvous time with the driver at Naphach Junction?"

"On Tuesday at 6 p.m."

"Does he know how to get there?"

"Yes. He has been there more than once."

"Did you ask him?"

"No."

"OK. Check it."

Then, the conversations got more challenging.

"Did you check that the training field in Kule is 'secured' for your training?"

"Yes. I talked with the officer in charge and he said that he secured the field."

"But are you sure that nobody else is going to be training in that field at the same time?"

"Yes. The officer in charge said so."

"Did you check it?"

"No."

"Ok, then check it."

And later, another challenge.

"Did you check that Nimrod arrives tonight?"

"Yes. I spoke with him."

"Are you sure he will arrive?"

"Yes. He is very reliable."

Giora leans towards me and raises his voice, "Are you sure?"

And I, experience again how my dad used to ask me, "How much is seven times eight?" And as I replied, "56," he would shout at me with his thunderous voice, "Are you sure?" And I would feel shaken. It took me years to be able to stand up against the frontal "assault" without flinching

I handle the verbal assault from Giora in the same way I reacted to my dad's questions. I block out any doubt, not even a split second of indecision, and answer with complete certainty, "Yes!"

Giora is content that I have reached the point of unshakable confidence, but can't help challenge it with, "Only death is certain!"

I feel as if the rug is pulled out from under my feet, I smile to myself and think, *You won again, Giora.*

I remember that I behaved that same way with my own children. I carried the pattern of my relationships with my dad over to my children, continuously quizzing them, "Are you sure?"—Maybe it's time to break that pattern?

Giora complimented me on my leadership. As Israel Independence Day approached, he asked me, "Do you want to be recognized by the President of Israel as one of the country's exceptional soldiers?"

I looked at him and said, "Giora, drop it. Just let me do operational missions."

We both laughed.

As a commander I remembered well the relationships we had as soldiers with our commander and the difficulties we faced together. We came to realize, the hard way, that we could only count on ourselves during the most demanding hours of operation. To know the details of the trail and to make the right operational decisions. It took us time to allow ourselves to see that our commander could make mistakes. This unthinkable thought was expressed out loud only once and then silenced. But as soon as we had spoken the thought, it surfaced more frequently and we began taking responsibility for navigating the trail and performing the missions. In the past it never would have occurred to me that a commander in the Unit could fall short. It did not make sense. Impossible. But one occurrence followed another, and the impossibility turned into "maybe." We kept suppressing the thought and told ourselves, "No, it isn't possible." But there was a disconnect. Something wasn't right. We saw our commander's undeniably impressive athletic abilities during training on the running trails. But we also experienced his limited ability to function adequately during the long freezing nights of hard work. Apparently, his athletic capabilities worked well only in the practice arena.

The disparity between the two enabled us, eventually, to see and acknowledge the problem clearly for the first time. Each area—operations and athletic practice—had different requirements, and our commander excelled in only one of them. That understanding allowed us to adjust and take measures to deal with our predicament as best we could. We understood that we had to step up during the long arduous hours of field operations. As a crew with many missions under our belt, we knew that we had to take responsibility for the countless details ensuring success and avoiding an operational failure.

During the first few months of our training in the Unit, we did not notice a thing. We existed in a daze of continual physical effort and mental exhaustion. In discussions we had with soldiers from other crews, we discovered that we were being shaped into a tougher, stronger crew. The number of miles we were demanded to walk were a bit farther. Navigation trails were longer than average by a few miles. The demands for "a bit more'" every day made the training week more challenging for us. There were times when we did not accomplish all the goals of the training missions, but the high standard demanded of us month after month built us into a much stronger crew.

Occasions such as the time Ovad and I went in search for the lost jeep while the commander slept, transformed the regular complaints of a soldier into a dawning understanding about the gap that was building up between us and our commander. Over time, the gap expanded.

The expulsion of Avner from our crew for no good reason added to the general feeling that something was awry. We knew something was wrong, but we did not dare to voice it. We were afraid and always chose to go ahead and perform what was asked of us instead of stopping to question, "What is going on here?"

For months, we would argue among ourselves whether there was a problem. The crew was split into two factions. As we accumulated operational experience, we came to trust our commander less and less. The argument that we were upset with him because he demanded too much from us lost power. We saw how, at more and more junctures, we made the decisions and he simply came along for the ride. Whenever there was a doubt about the next steps to take along the trail, he consulted with us and decided according to our opinion. When there was a need to make an operational decision about the sequence of performing tasks, we went with our own choice, sometimes contrary to his orders. We did this to avoid mistakes. It became very clear that the crew were the professionals and the commander was merely "along for the ride." We talked among ourselves and all of us decided to acknowledge this reality.

The trust gap widened. At the end of an exercise in Kule, where the commander made a wrong decision, we returned to the Unit and Yadi, who was always one of the calming voices in the crew, joined with the rest of us in declaring the commander unfit.

After Shai froze to death, we demanded that Giora find a new, experienced commander to lead us. Fortunately, Zvika became our lead commander for the next mission. We were very pleased that a practiced, proficient commander was now in charge and felt relief and a new sense of confidence. We knew that from now on an experienced, capable leader would be joining us in the field.

Decades later, after our official discharge from the IDF, I was surprised to hear that our former commander joined another crew on some missions and contributed positively, in an impressive way, to helping a less experienced commander.

Maalot Massacre

How is the hammer of the whole earth cut asunder and broken!
(Jeremiah, Ch. 50, V. 23)

May 1974. A terrorist attack on school children in Maalot. Hostages are taken. Utter horror and shock.

Early in the morning a helicopter lands near the field where we are undergoing mortar training on the West Bank. The soldiers from the training base watch us with admiration as we take off to fly to the Unit base. We quickly organize our gear to travel immediately to Maalot. My crewmates board the chopper and take off. I am sick with diarrhea and a high fever. The commander orders me to see the doctor, and the doctor instructs me to lie in bed, drink water and take aspirin. I am so exhausted that I do not even argue. Get into bed, drink some water and fall asleep for a few hours. When I wake up, I still feel very weak but the fever is much lower. I ask myself, *Where is everybody?* then remember, *They are on their way to Maalot. How do I get there?*

Although I am still recovering from my illness, I see a truck on the pavement in front of me. I talk with the driver. He tells me that he is leaving for Maalot in a couple of minutes. I quickly fetch my rifle and army belt, as well as an apple, some water, and a dry piece of bread.

Dad used to say that apples are good for curing diarrhea. In the camps, if the diarrhea continued, you would swallow a piece of coal to stop the runs and save your life.

I also take a few aspirins with me to be on the safe side. Climb into the truck. Drink from the canteen, swallow the aspirin and fall asleep. We are moving. My fever climbs. Heading north, the potholes in the road shake me, but I fall asleep anyway.

I day-dream.

In the Buna Manoviz concentration camp, Dad fell ill and went to the camp hospital. High fever. Day after day. On the third night, he felt somebody throw him to the bench on the other side of the aisle. The following morning the Doctor told him, "You have to get out of here. If I had not thrown you to the other side, they would have marked your arm with "L," meaning "Leiche" (German for "corpse"). Get out of here and do what you have to do to survive. Here you will die."

I wake up to the sickening foul scents of Acre Gulf, north of Haifa, coming from the perfume factory. We continue north through the industrial zone. I fall asleep.

I can hear the roar of the truck engine through the haze of my fever as the truck climbs from the Mediterranean Sea up into the mountains. Swallow another aspirin. "Arbeit Macht Frei." Work sets you free. So does aspirin. The air cools down. We arrive in Maalot.

A squad of three terrorists have captured about a hundred high school students from Zefat who were on an annual trip. They are holding them captive inside the Nativ Meir School in Maalot. Moshe Dayan, the Security Minister, Mota Gur, the IDF chief of staff, and Refael Eitan, the general in charge of the north, are all on the scene.

It is afternoon. We set up near the school. I drink again, take another aspirin and join the briefing. We are the force designated to enter the building by going up the stairs. We will not be required to engage the terrorists, I think since there are snipers and other crews that will be entering the building through the windows. By the time we arrive on the scene, the whole event will likely be over.

We move forward in a long line behind the school wall so the terrorists will not spot us. I stand behind Amnon, and there are a few soldiers from the crew of Zvika Livne in front of us. Many more soldiers are behind me. We lean against the wall and wait for the signal. The sun moves westward. The order comes to get ready. We are about to go in.

This is the first time a terrorist squad has held such a large group of school children hostage and they are demanding a ransom for their release. It is a new situation that none of our commanders have experienced before. The government decided to bring in its best soldiers because nobody knows what to do in such a situation. We do not know either. Everyone hopes that, with the versatility and knowledge gained from our combat experience in the Unit, we will be able to handle the situation.

I stretch. Tense and release my left palm holding the Uzi. Do the same with my right palm that holds the cartridge housing. My right finger remains on the guard around the trigger. The safety guard is set on single shot. My pulse is steady as I draw a deep breath. The school wall is on my right and a line of cypress trees is on my left.

I visualize myself running up the stairs. One flight, turn left. Another flight, right. Rush to the classroom entrance at the end of the corridor. Move the machine gun to the center of my body as I enter the classroom. Spot the enemy, get him in my sight, fire, strike my target. Make sure he's dead, and "get back" to our line. Chase away a pesky fly with sharp motion of my head. Time passes and we are finally given the message, "Soon."

Tension builds as soldiers up and down the line are alerted. I take a small step forward to close the gap between me and the next soldier without getting too close. Check my boots and lean over to tie the laces more tightly. Move my left hand to the pouch to check that I have enough cartridges. The hand grenade is in place on my belt. Right hand to check my remaining pouches. Let it drop to my side.

What's the status of the hostage children on the second floor? I suppress that thought. Visualize myself running into the classroom. Aim for the center mass of the terrorist and fire. Focus on the task at hand and chase away any other thoughts.

"We are about to move."

The order for "ready" passes through the line and causes everyone to tense with heightened alertness. I hold my Uzi in my palms ready to use it. Roll my shoulders. Breathe. Take one step forward.

The sounds of weapons fire. The line of soldiers moves a few yards ahead and then stops. Takes a step backward. I see the canteens on Amnon's army belt ahead of me. I hear intense bursts of gunfire.

TRRRRRRRRR....

Amnon and I run with all our might to get inside.

"Lighter than eagles. More Powerful than lions."

TRRRRRRRRR.... TRRRRR .... TRRRRRRRRR....

Run as fast as I can. TRRR... TRRR... TRRR...

Run through the thick-blinding-choking smoke and up the stairs.

The sounds of weapons firing continue. TRRR... TRRR... TRRR...

Run full out. TRRR... TRRR... TRRR...

Through the smoke and up to the second floor. TRRR... TRR... TRR...

Turn right to the corridor and race towards the classroom at the end of the hall. TRRR... TRRR... TRRR...

Amnon and Baruch arrive a split second ahead of me. Maddening dash.

They are already in the room. I enter and position myself on the left near the door. My Uzi is ready to shoot as my eyes sweep the room at once searching for the terrorists. Complete silence.

As I enter, Amnon shouts, **"Don't shoot. We killed them."**

Heavy smell of shooting smoke. I stand by the wall. The silence after the intense gunfire is deafening. I look around the room intently, ready to shoot in case a threat arises. My eyes and the Uzi barrel are searching attentively.

Clean.

From the moment of entry until now less than three minutes have passed. A lifetime compressed. I now have time to look around.

Survivors are groaning on the floor. The three of us search the room to make sure no terrorist remains alive. None of the terrorists are alive. Some of the bodies on the floor are moving. Groaning. For a split second I am shocked at the picture of so many wounded bodies on the floor. I unload the Uzi and verify that the weapon is empty and no longer a danger.

This looks like a scene in a gas chamber from the Holocaust.

Many wounded children in the class room. The outcry of anguish that is about to erupt from me is transformed into an intense need to rescue the wounded. I grab the nearest hurt boy in my arms and hastily run down the stairs toward the school entrance. I slow my pace so as not to hurt the wounded child I carry.

"Bring stretchers!" we shout. "Many stretchers!"

Paramedics arrive. Run back up to the classroom, as quickly as I can, to evacuate more wounded. Other people join us. We run again and again with the wounded in our arms and carry them to the waiting stretchers. Evacuate the injured children from the school and

bring them to the long line of ambulances. Return to that nightmarish classroom over and over. Each time I enter that room, images of
the stacks of naked bodies taken from the camp gas chambers flash
through my mind. I am functioning on autopilot. I have no time to
"see" the devastation around me. Just bring the children out of the
room and down the stairs one at a time. Quick. Quick.

The medical teams attend to the children and place them in the
waiting ambulances for transportation to the nearby hospital. I see a
newspaper photographer taking pictures and think: *What for?*

Enter the empty, blood-soaked room one last time, take a look,
and go down the stairs. The evacuation of the wounded has ended.

Moshe Dayan leaves. We load our gear on the truck getting ready
to travel south. The intense activity is replaced by a strong sense of
failure. We get on the truck and each take our customary spot. The
truck moves south while we feel weighted down by our failure to
prevent casualties.

We ask ourselves, *How could we have handled this differently?
How could we have taken the terrorists by surprise? How could we have
arrived at an unfamiliar place and taken control of the situation immediately ensuring no hostage would get hurt?*

The following morning, we conduct a thorough investigation of
the operation and start developing plans and practices to take control of similar situations in the future differently by devising other
ways to enter such rooms.

In parallel we assisted in building a new IDF unit that would
specialize in dealing with hostage situations. The next time we would
be better prepared.

Late at night my thoughts go back to that classroom and the
scene of wounded and dead bodies. Horrified. Happy that Dad and
Mom managed to escape a similar fate.

The following morning our crew resumes its intense training.

During the terrorist attack in Maalot 22 children from the Zefat high school were murdered in that classroom. A three-man terrorist squad crossed over the border with Lebanon. On the way to the school, they also murdered a family in Maalot. They attacked the school and took about hundred children hostage. Most of the teachers and a few children managed to escape but the majority of the children were trapped in the classroom.

In the terrorist attack on Maalot, twenty-seven Israelis were murdered and the three terrorist killed.

Biking to Eilat—3

Cool morning in Faran. Lonny went out two hours ago to bicycle the 20 miles he did not get to finish yesterday. A policeman ordered him to stop riding and get in the escort car since it was getting dark. Today the trail is short. Only 60 miles to Eilat. Lonny is unwilling to miss out on even one mile of the journey. Past scars. He woke up early and caught a ride north to the point where he had stopped yesterday when we had already arrived at Faran. We get ready for our day's ride today after eating a wonderful breakfast that Metula, Ovad's wife, prepared for us. We meet Lonny at the entrance to the Moshav and climb the hills together on our bicycles. We ride down to Kushi with a light morning wind at our backs. Less than 55 miles and the whole day ahead of us. We peddle and talk casually as we ride side by side. The 60-mile ride feels like a vacation.

We pass Kibutz Eylot and enjoy the last downhill ride of our trip as we enter Eilat. Not yet noon. The amount of talking has increased 500% in the last few minutes. The beauty of the blue sea, the red mountains, the runway and the hotels are a welcome sight for me. My muscle aches are subsiding as I embrace the elation of conquering this challenge. A saddle sore on my inner thigh makes its presence felt after the hours of friction with the bicycle seat. We get off the bicycles and follow Lonny to park them in the hotel's parking lot and enter the lobby.

The receptionist asks, "Where are you coming from?"

"Tel Aviv."

"Did you leave a week ago?"

"Yesterday."

"Whaaat?"

I feel a profound sense of satisfaction.

Another few limping steps and I am in my hotel room and standing under the shower. Put on a white robe and sit outside on the balcony, looking out towards Akaba across the Red Sea.

Breathe.

Coming Home

Friday evening. I get out off the car at the entrance to my kibbutz. Look at the familiar ascent to the gate at the top of the hill. Cross the highway and start walking. Look behind me and see the loess-covered road to the orchard that was uprooted years ago. Now all is quiet. Can envision the clouds of dust that follow jeeps speeding down this road. No dust clouds or jeeps right now. All of the farming equipment parked under the tin roofs of outbuildings. Darkness. No car traveling on the highway. Walking energetically with a sense of contentment while clearly feeling my sore muscles. This feeling is an integral memory from my IDF service. My feet hurt. Nothing out of the ordinary. Ascend the path. Imagine meeting tomorrow afternoon with Lotan, Edna and Ofira as we sit on the grass in front of Lotan's room with vanilla coffee flavored ice cream in our cold coffee mugs and chat.

Carrying my bag still hurts from the weight that tore muscles of my shoulders. The air invigorates me. The familiar dry air of the desert is mixed with the gentle smells of pines. As I walk down the path, the smell thickens and engulfs me. Home. The scent of pine enriched with eucalyptus follows me downhill. Intoxicating. The scent of freedom. The rich aroma mingles and swirls with the gentle gusts of wind. The sounds of classical music trickles from a nearby window. The fragrant scents and the music envelop me and fill me with a sense of peace and contentment that spreads throughout my body. The joy of being home and among my friends fills my heart. I am weeping. My pains dissolve. Breathe deeply. Savor each moment of existence. Am not in a hurry yet widen my steps.

Breathe in freedom.

Photos

In the unit near our rooms, About two years after recruitment.
It is common to be in civilian dress in the Unit.
From left to right: Lony, Dany, Yadi, Shai, Adi, Doron, Ovad.

On the left, Yadi, shaking hands with me.

*Navigation in the Negev Desert—in the south part of Israel—
from left to right: Adi, Ben-Ami, Lonny, Amnon.*

*Rest during training in the Golan Heights—
from left to right: Lonny, Ouri, Ben-Ami.*

Training at sea—
From left to right: Shai, Yadi, Doron, Adi, Hagai, Ben-Ami, Ouri, Ovad, Amnon.

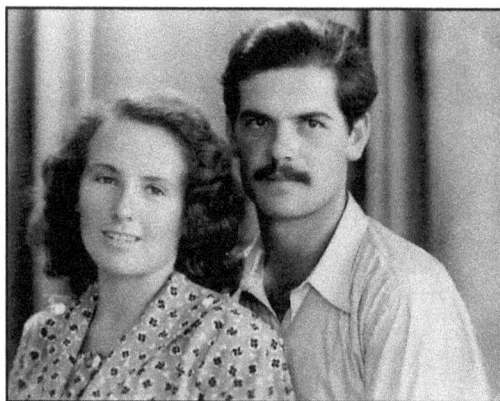

My mom, Shari, and dad, Yehoshua.

Addition to the Second Publication

After I finished writing *Along the Trail*, a miracle happened to our family. Following a "roots trip," that began with research spear-headed by my brother Asaf and culminated in a journey to Poland and Germany by my siblings, my father and me, my sister Iris began writing our family history. She had a question about the details regarding my family's time entering the Krakow Ghetto and request-ed information about the Yehoshua Libliech family from the Shoah Museum in Washington, DC. A few weeks later, Iris received about 50 detailed documents about Sheindel Leah, the eldest sister of my Dad Yehoshua, including pictures of documents prepared by the Germans during World War II and a few years thereafter. My sister was stunned and mailed the documents to Israel, assuming there was some kind of mistake.

When my brother Assaf inspected the documents, he discovered that Sheindel had not only survived the Holocaust, but had also im-migrated to Israel. He visited the Israeli Interior Office and received confirmation that Sheindel and her husband had lived in Haifa and passed away during the 1970s. After making an inquiry with Hevre Kadisha, the organization in charge of burials in Israel, he found out that Sheindel had two children and was buried in Haifa. Following this discovery, we were able to track down Sheindel's son and daugh-ter and meet them.

Unfortunately, Yehoshua and Sheindel were not re-united after the war despite the fact that both lived near each other in Israel for many years. Both thought that they were the sole survivors from their family. The meeting of Yehoshua with the children of his beloved sister Sheindel, who was the subject of many of his writings, was extremely emotional and documented in a short movie created and produced by Alon Degani from Kibbutz Mishmar Hanegev. The reunion was filmed live and shown on Channel 1 in Israel.

The movie, available on YouTube, can be accessed online by entering *https://www.youtube.com/watch?v=pKR_S7Xw92w* or the film's title, *Discovering Family after 70 years*, in the search box.

After the publication of the first edition of *Kol Haderech*, I received many warm responses from people from many parts of Israeli society—mothers and fathers of IDF soldiers, IDF soldiers from many different units, educational professionals, Unit veterans, literature lovers, Reuven (Ruvi) Rivlin, Israel former president, and boys and girls preparing for their service in the IDF. I had the honor of being interviewed on the radio by Anat Dolev on channel B.

I received a prize for creative military literature. Below is a picture from the awards ceremony.

One of the surprising and most heart-warming responses arrived from Israel Genosov, who with outstanding generosity offered to fund the publication of the second edition of *Kol Haderech*, and I am deeply thankful to Israel.

Glossary

Aba – The Hebrew word for Dad.

Natan Alterman (1910 -1970) – An Israeli poet, playwright, journalist, and translator, highly influential in Socialist Zionist politics, both before and after the establishment of the State of Israel.

Auschwitz – Auschwitz-Birkenau – The largest Nazi extermination camp complex, located in Poland, where hundreds of thousands of people were murdered during World War II and the Holocaust.

Bakibbutz – A weekly magazine that described the lives and events in the kibutzes all over Israel.

Bar-Kokhba – A Jewish military leader who led a revolt against the Roman Empire in 132 BCE. Its failure was a disaster for the Jewish people. A substantial number of Jews living in Israel were murdered or enslaved by the Romans.

Beer Sheva – The largest city in the Negev Desert of southern Israel, often referred to as the "Capital of the Negev."

Beit Borchov – A building situated in Mishmar Hanegev in the memory of a leading thinker of the Kibbutz movement.

Ben Gurion Airport – Located on the northern outskirts of the city of Lod, between Jerusalem and Tel Aviv, it is the main Israeli airport. In 1973, it was renamed from Lod Airport after Israel's first prime minister, David Ben-Gurion.

Buchenwald – A Nazi extermination camp near the city of Weimar in Germany.

Buna Monowitz – A subcamp of the Auschwitz concentration camp complex in Poland.

Matti Caspi – A talented composer, musician, singer, arranger, and lyricist known and loved by most Israelies for the last four decades. Born in 1949, he is considered one of Israel's top popular musicians.

Choshech Mizraim – A very deep night darkness where one barely can see its own palms – the darkest of dark. Choshech Mizraim is used for the darkest darkness. It also refers to the dark period in Jewish history when the Israelites spent 400 years in Egypt, mostly as slaves, before Moses led them out of Egypt.

Eli Cohen – An Israeli spy who infiltrated the highest circles of the Syrian government and provided critical information prior to the Six Day War. He was caught and hanged in Damescus in 1965. The Netflix drama series, *The Spy*, starring Sasha Baron Cohen, tells his amazing story.

Moshe Dayan – Grew up in Moshav Nahalal, the first Moshav in Israel, and became the minister of defense during the Six Days War. With his black eye patch that became his trademark, he cut a dashing appearance and became an international symbol of the fight for Israeli independence.

Dead Sea – A salt lake located south of the Sea of Galilee and fed by the Jordan River. Its shore is the lowest dry land point on Earth – 1,412 feet below sea level. It got its name because its high salt content – around 35% - makes it impossible for fish and aquatic plants to live in it.

Menachem Digli – A former commander of the Unit who was in charge of information gathering for the IDF military intelligence before the Yom Kippur War.

Double steps – A way to measure distance walked by taking two steps forward, one with the right foot, and the other with the left. Each navigating soldier knows how many "double steps" he had to walk to cover 100 yards.

Druze – A religious and ethnic minority of Israel and the surrounding countries, with a distinctive culture. Druz live in Israel, Syria, and Jordan are known as very brave soldiers.

Death March – A forced march of hundreds of thousands of concentration camp prisoners toward the end of World War II instigated by the Nazis to escape the advancing Russian Army. During these Death Marches the majority of the prisoners perished.

Levi Eshkol – The Israeli Prime Minister during the Six Days War.

EZEL – A Jewish paramilitary resistance organization that operated in Palestine after World War II. It was one of three underground movements that fought the British occupation forces before the Israeli Independence war of 1948.

Galilee – The northern region of Israel characterized by high mountains and wide valleys, where a sufficient amount of precipitation allows the cultivation of a variety of agricultural crops and keeps land is green most of the year.

Galilee Finger – The northernmost portion of Israel.

Golani – An Israeli infantry unit like the paratroopers. Both the paratroopers and Golani compete who is better.

Shmaryahu Gutman – A famous Israeli archeologist who led a significant portion of the excavations at Masada.

Hanoar Haoved and **Machanot Haolim** – Two youth movement organizations that work with 12- to 18-years-old teenagers to encourage them to eventually choose to live in a kibbutz. The Hanoar Haoved movement is larger and engages with teenagers across Israel while Machanot Haolim is quite small and involves mostly youths from higher socio-economic, urban areas.

HaKibbutz Hameuhad – One of four Kibbutzim movements that was aligned with one of the ruling parties in Israel. Two other kibbutz movements were Hashomer Hazair and Ichud Hakibuzim. The forth movement is religious. Although, kibbutzim comprised less than 2 percent of Israel's population, they are responsible for 40 percent of its agricultural output and contribute substantially to manufacturing as well. They also played an important role in the building and defending Israel during its formative years.

Herod – A king of Judea, also known as Herod the Great, who ruled with Roman support from 30 to 4 or 1 BCE. He is considered as the greatest builder at the time. His projects included the expansion of the Temple Mount in Jerusalem and the construction of the fortress at Masada during the historic times of Israel.

Herodion Mountain – A cone-shaped hill located south of Jerusalem in the West Bank, the remains of the palace fortress Herod the Great built there.

Hermon Mountain – The highest mountain in the north of Israel (9,232 feet above sea level). The Hermon is located on the border of Israel, Syria and Lebanon.

Ayin Hillel – The pen name of Hillel Omer (1926 - 1990) an Israeli poet and children's author.

IDF – Israeli Defense Force – The umbrella name for all military organizations in Israel.

Kadoorie School – An agricultural schools endowed by British Jewish philanthropist, Ellis Kadoorie, and founded in 1933 during the British Mandate of Palestine. Located in the Galilee near Kibbutz Beit Keshet, its most famous student was Yitzhak Rabin.

Kastina – A major traffic junction on the road from the center of Israel to the South. From there, the main road heads to Beer Sheva, the largest city in the Negev Desert, and the secondary road leads to Ashkelon, the last city on the way to the Sinai Peninsula.

Kibbutz Mishmar Hanegev – Located 12.5 miles north of Beer Sheva, it is the kibbutz where I grew up.

King David – A ruler of the kingdoms of Israel and Judeah, father of King Solomon. David won a famous fight with Goliath, the Philistine champion and, as king, conquered significant amount of territory.

Kus emek – An obscene curse word, originally from Arabic, meaning "Your mother's pussy." it is shorthand for cursing the moment someone came out of their mother's womb, i.e., the fact that they were even born.

LaMerhav – A political magazine associated with Ahdut HaAvoda, a forerunner of the modern-day Israeli Labor Party. First published in December of 1954, it was merged into *Davar*, another Labor newspaper, at the end of May of 1971.

Amiram Levin – A retired general in the IDF former commander of Sayeret Matkal who was regarded as one of the most couragous soldiers in the history of the Unit.

Levinstein Center – The main rehabilitation center for wounded soldiers and civilians in Israel.

Lod Airport – (see **Ben Gurion Airport**)

Lot – A nephew of Abraham in the Bible, who resided in Sodom and was considered the most righteous person in that city.

Luf – Kosher ground canned meat of poor quality, akin to SPAM, used as primary combat rations by the IDf. Many Israeli soldier claim that Luf uses all parts of the cow.

Machanot Haolim (see **Hanoar Haoved**)

Mag – The standard machine gun used in the IDF.

Masada – The last strong hold of the Jews during the rebellion against the Romans in 70 AC. A few hundred warriors held the fortress against the surrounding Roman legions and killed themselves rather than allowing themselves to be taken prisoners. One of the slogans ingrained in the recent Israeli culture is "Masada will never fall again." Some IDF soldiers pledge allegiance to Israel on Masada.

Golda Meir – An Israeli stateswoman, politician, teacher, and kibbutznik, who replaced Levi Eshkol as Israeli prime minister and served during the Yom Kippur War.

Migda – A farm that belonged to Kibbutz Mishmar Hanegev, where irrigated crops were planted.

Nablus – An Arabic word our crew members adapted as an expression meaning "to huddle tight and close to one another." It is also the name of a large city on the northern part of the West Bank.

Nafach – The remains of a village destroyed in the aftermath of the 1967 War. One four-room house had its roof still intact, and we used to eat and sleep there during our weekly training in the Golan.

Navigation point – A spot on the navigation trail. The navigator has to describe that point to ensure he was in that point during the day/night navigation trail.

Nord airplane – An old piston plane from WW2 that used to be a carrier and became the only plane for training paratroopers.

Pankieviz Pharmacy – The Pankieviz Pharmacy was on the border of the Krakaw Gettho and Plaz Visgodi which was the place were Jews transported to the extermination camps. Pankieviz was a Polish pharmacist who enabled smuggling of necessities to the ghetto to save lives.

Yzhak Rabin – Former Israeli prime minister who studied in Kadoorie, became the IDF chief of staff. Rabin was assassinated by a right wing extremist.

Red Sea – The sea in the south of Israel connecting Israel to Africa and Asia through the Indian Ocean. Eilat is the port of Israel to the Red Sea.

Moaviat Ruth – A woman who came from Moav to Judea and married Boaz. Ruth is the grand mother of King David.

Sami Bourekas – Sami Bourekas is a famous shop of savory pastry shops. Bourekas are pastry made with filo dough and stuffed with cheese, spinach, potatoes or meat.

Sayeret Matkal – The elite intelligence gathering force, also known as the Unit, in Israel and the IDF.

Sayeret Golani – The elite unit of Golani.

Dany Senesh – Born January 7, 1953, died June 24, 1974. Born and raised in Moshav Ben Ami. Became a combat soldier in Sayeret Matkal and then a combat soldier in Sayeret Golani. Was killed while leading the offense against terrorists in Naharia.

Shai Shacham – Born April 19, 1952, died November 2, 1973. Born and raised in Kibbotz Kabri. Became a combat soldier in Sayeret Matkal. Froze to death on a rescue mission in the Hermon Mountains.

Shalhin – Crop grown by irrigation with sprinklers and droppers. In the kibbutz we used to grow crops such as potatoes, cotton, carrot, and wheat.

Sheik grave – A grave of a holy Islamic person usually placed at the top of a hill. The tomb of a holy Muslim person. Israel is spotted with thousands of Sheik graves, most of them located on top of hills.

Shoham – A town between Jerusalem and Tel Aviv.

Siftach – An Arab word meaning "beginning" or "opening." Usually, merchants in the market are eager to sell the first item, and they give a discount to the buyer and make a "Siftach," a "blessed opening" for the day.

Six-Day War – Also known as the 1967 War, it took place in June 1967 when Syria, Jordan and Egypt all attacked Israel. The war ended more quickly than expected, with Israel victorious and conquering land from all three countries, notably the Golan Heights from Syria, the Sinai Peninsula from Egypt and the West Bank from Jordan.

Soroka Medical Center – As the general hospital of Beer Sheva, Israel, and central hospital of the region, it provides medical services to approximately one million residents of the southern part of Israel.

Suez Canal, El Arish, Bardawill, Refidim, Tasa, Baluza, North Sinai marshes, Kadesh Barne Oasis, Sharem Al Sheik – All are points of significance for IDF soldiers who served in the Sinai desert.

Syrian-African Rift – A fault that starts in Tanzania and runs north to Turkey. In Israel, the Red Sea, Arava Valley, the Dead Sea, the Sea of Galilee and the Hermon Mountain all lie within its territory.

Tel Grar – An important biblical town on Nachal Grar Wadi, near Kibbutz Mishmar Hanegev. Tel Grar is mentioned several times in the Bible as the ruling city of the King of Grar in ancient times.

Tolar – A light cannon placed on a jeep enabling infantry forces to combat enemy tanks.

Yalla – Meaning "Hurry up" or "Let's go" in Arabic, it is used Israeli slang to encourage someone to do something.

Yom Kippur – Literally "day of atonement," it is the most sacred day in the Jewish tradition, culminating a ten-day period of penitence that begins with Rosh Hashanah, the Jewish New Year. On Yom Kippur, no one in Israel works, and there is no radio nor television.

Yom Kippur War – Also known as the October War, the 1973 Arab–Israeli War, or the Fourth Arab–Israeli War, it took place from October 6 to 25, 1973 between Israel and a coalition of Arab states led by Egypt and Syria with large number of casualties on both sides. When Israel started to defeat its attackers, a ceasefire officially ended the war.

Flavius Yosephus – Born in Jerusalem, he lived from 36 to 100 and was one of the leading army commanders in the rebellion against the Romans in the First Jewish-Roman War (70 AC). Surrendering to Roman forces after the six-week siege of Jotapata, he later became the foremost historian of this period.

Giora Zorea – The commander of Sayeret Matkal (the Unit) during the Yom Kipur War.

Acknowledgments

I want to thank all of the many people that contributed to the completion of this book.

Hadar, my wife, who challenged and supported me in this endeavor, and edited the book. You helped me be more precise in my language and more communicative.

My children Elia, Etay and Efrat, whose curiosity and many questions spurred my desire to write this book.

My dad, Yehoshua Tzafrir, who nurtured me and my memories with his words, poems and stories.

My mom, Shari, who raised me and took care incessantly.

My brother, Assaf and, sisters Ora and Iris, who supported and encouraged me along this trail and offered their comments.

Ester and Shmuel Meoded, who supported and encouraged me.

Ilan Baruch, who was sensitive and encouraging at the beginning of this road and instilled in me the confidence to continue.

Amitay Korn, for asking probing, direct questions that encouraged me to dig deeper.

Ezra Zahor, for your wonderful pictures.

Ruty Kaplan, for your patience and your photographs.

Roni Ashernizky, for teaching me the language of images and photography.

Shlomo Adler, for weaving Bible verses through my story wisely and beautifully.

Noam Verner, for drawing so skillfully the maps.

Giora Zorea, for fact-checking and correcting and helping me see the commander-crew relationships from another angle.

Ehud Barak, who with gentle insistence complimented my efforts and shared his reading enjoyment.

Amiram Levin—your words energized me.

Muki Bezer for sharing the pains of long-distance writing.

Alik Ron for encouraging me to continue and create.

Heartfelt thanks to my good friends Eran Kopel, Zur Tene, Shahaf Aaude, Uri Ganani, Hagit Porat, Dania Galili, Yael Zimler, Roni Krayn, Roby Kaplan, Amir Ezrahi, Yalon Lotan, and many others who read, commented, asked questions and always encouraged me.

My friends from the Bracha's circle who gave me space for experimenting and expanding my writing.

My friends from the Kolot circle, who warmly supported me, especially Sarit Rothschild, who loved, commented, and enlightened with a sensitive touch.

Noam, who edited professionally, mixing tenderness with precision.

Tamar, who cleaned up and tightened the Hebrew version.

The Senesh and Shaham families. Your sons, Dany Senesh and Shai Shaham are sources of inspiration to me and are deeply missed.

Israel Genosov, who generously enabled the printing of the second edition of the book.

Lisa Springer, who worked diligently on the English translation of my book.

Special thanks to my dearest crew members: Adi, Yadi, Avner, Ovad, Lony, Doron, Ben Ami, Hagai, Amnon, and Yehuda. The experiences were created with you. What I have written is to you and for you.

Addition to the English Version

Many thanks to Lisa Springer who worked with me for months attempting to understand the Israeli culture and correcting my English.

A whole-hearted thanks to Yehuda Inbar for his inspiring contribution that enabled the publication of this book in English. I am forever grateful for his generosity.

Gratitude and deep appreciation to Chris Angermann, who persistently raised the English version to new heights while connecting with it on a deep, intimate level.

Ouri Tsafrir, a business consultant, lives in Shoham, Israel with Hadar, his wife. They have three children ages 30, 28 and 23. Growing up on a kibbutz with two Holocaust survivor parents, whose entire families were murdered, deeply affected his life. After high school he joind Sayeret Matkal (the Unit), the most demanding IDF commando unit and became a combat soldier and a crew commander. Along the Trail, which took him 10 years to write, enabled him to clarify and work out some of his memories from that intense period.

Ouri lectures in high schools, Mechinot (pre-military education groups), various IDF units, and other groups that desire to hear firsthand about his experiences and the process of writing the book. He continues to keep in close touch with his crew members from the Unit and meets up with them frequently on trips or at social gatherings.